WHAT TO SAVE AND WHY

WHAT TO SAVE AND WHY

Identity, Authenticity, and the Ethics of Conservation

Erich Hatala Matthes

OXFORD
UNIVERSITY PRESS

Oxford University Press is a department of the University of Oxford. It furthers the University's objective of excellence in research, scholarship, and education by publishing worldwide. Oxford is a registered trade mark of Oxford University Press in the UK and certain other countries.

Published in the United States of America by Oxford University Press
198 Madison Avenue, New York, NY 10016, United States of America.

Library of Congress Cataloging-in-Publication Data
Names: Matthes, Erich Hatala, author.
Title: What to save and why : identity, authenticity, and the ethics
of conservation / Erich Hatala Matthes.
Description: New York, NY : Oxford University Press, 2024. |
Includes bibliographical references and index.
Identifiers: LCCN 2024013828 | ISBN 9780197744550 (hardback) |
ISBN 9780197744567 (epub)
Subjects: LCSH: Cultural property—Protection—Moral and ethical aspects.
Classification: LCC CC135.M38 2024 | DDC 363.6/9—dc23/eng/20240422
LC record available at https://lccn.loc.gov/2024013828

DOI: 10.1093/oso/9780197744550.001.0001

Printed by Sheridan Books, Inc., United States of America

For Etta

CONTENTS

Acknowledgments ix

1. Introduction 1
2. From Heirlooms to Tacos: What Should We Save? 16
3. From Wilderness to Pottery: Why Should We Save Things? 43
4. From Climate Change to Colonialism: From What Should We Save Things? 70
5. From Language Revitalization to Digital Replication: How Should We Save Things? 99
6. From Appropriation to Participation: By Whom Should Things Be Saved? 127
7. From Ourselves to Future Generations: For Whom Should We Save Things? 155

NOTES 179
REFERENCES 197
INDEX 221

ACKNOWLEDGMENTS

I'm grateful to many people who supported the writing of this book. I've been thinking on and off about the philosophical dimensions of conservation since graduate school, and I am grateful for the supportive intellectual community I found at Berkeley. Much of the literature that I engage with in this book I first taught in my courses at Wellesley, and I am grateful to my students for thoughtful discussion of these ideas. In particular, I want to thank the students who have taken Philosophy of Art, From Wilderness to Ruins, Who Owns the Past?, Authenticity, and Environmental Aesthetics. The opportunity to teach such a wide array of courses has been made possible by my supportive colleagues in Philosophy at Wellesley, to all of whom I owe thanks. Special thanks to my friend and colleague Julie Walsh for her feedback on an early draft of the manuscript. I am also grateful to Wellesley College for a semester of research leave, which allowed me to draft the bulk of the book.

I've benefited from the opportunity to present work on conservation ethics at a number of venues, including the Getty Villa, the MIT School of Architecture, and the Bard Graduate Center, and I am grateful to hosts, colleagues, and audiences for their feedback. Special thanks to Ivan Gaskell for multiple exchanges that have pointed me in the direction of valuable sources and informed my understanding of conservation across disciplines. In a similar vein, I thank Elizabeth Marlowe and the other participants in her heritage studies reading group, which has been a fantastic source of inspiration and interdisciplinary learning.

Thanks are also due to the American Society for Aesthetics (ASA) and the excellent academic community it has fostered. This book has been informed by the scholarship of many ASA members, most of whom I have also had the opportunity to get to know personally over the years. Much of that scholarship isn't explicitly *about* conservation, but its value cuts across topical boundaries and offers insights that can shape our understanding of many different philosophical problems. It's hard to imagine what my career in philosophy would have been like without the ASA crew, a group that grows with every meeting I attend.

Tremendous thanks are owed to my editor at Oxford University Press, Lucy Randall, for her support of this project and her productive feedback throughout the writing process—Lucy is wise, and thoughtful, and a pleasure to work with. Thanks, too, to everyone on the OUP production team who has helped bring this book into existence.

Everything in life would be harder and a lot less fun without the love and support of Jackie Hatala Matthes—thank you for being my constant companion on this wild ride. Thanks to my parents, Elisa Matthes and Richard Matthes (1944–2012), for fostering my love of learning, and to my sister, the poet Sarah Matthes, who is always an inspiration. Finally, thank you to my daughter, Etta Matthes, to whom I dedicate this book. Everything in the world that I want to save, I want to save for you.

I | INTRODUCTION

Time turns us all into conservationists. Not time itself, perhaps, but the forces that time drags with it. Wear, tear, degradation, decay. Abandonment, neglect, disuse, dereliction. *Sic transit gloria mundi*. I first encountered those then-inscrutable words across the marquee of the independent movie theater I frequented in college, on the day it closed its doors for good. Black plastic letters stark against a glowing white background that once listed new releases and cult classics. No Latinist, I looked it up: *Thus passes the glory of the world*.

If we want to save the things we cherish from time's ravages, then we need to preserve them, conserve them. Easier said than done. After all, there's so much worth saving, and the list grows with every passing day as we innovate and imagine, create new artforms and inventions, generate novel customs and practices. Meanwhile, the things we care about face threats that would accelerate their loss, from climate disturbances to war, pressing the need for conservation upon us with ever greater urgency.

Concerns about conservation crop up in contexts that range from major art museums to the corner of your attic, from regional language communities to national parks. They confront us over kitchen-table conversation and on frontpage news. In May of 2022, Kim Kardashian wore a Marilyn Monroe dress to the Met Gala and the internet went absolutely bonkers about it. It was the social media controversy du jour, and major news publications continued running articles on the "scandal" for weeks. Kim damaged the dress! Or did she? Should she have been allowed to wear it? Should it belong to a company like Ripley's Believe It or Not (the owner who loaned it to Kardashian) or is its proper place in a museum?[1] Well, it's just a dress, you might say. But the word "just" is often a rhetorical last resort, a weapon we grope for when we've run out of arguments. From heirlooms to culinary traditions, classic cars to ancient trees, the things we care about saving are rarely *just* what they seem to be.

The significance of what we choose to preserve often sits at the heart of culture war controversies. The walls of George Washington High School in San Francisco are adorned with a 1936 mural by Russian-American painter Victor Arnautoff. Arnautoff, a member of the Communist Party, wanted to highlight the ways that enslavement and colonialism formed the basis of American economic success, and the panels composing his *Life of Washington* depict enslaved Africans and a dead Native person. Despite the affinity between Arnautoff's aims and the contemporary progressive push to teach students about America's history of injustice, the mural has been a periodic source of tension for decades, most recently in 2019 when the San Francisco school board voted to *destroy* the mural, with one board member framing the decision as a form of reparations.[2] The concern is that it is unjust to subject Black and Indigenous students to these violent images as they try to simply go about their school day. The school board ultimately changed their decision to destroy the mural in favor of covering it, and a newly constituted board (after a contentious recall vote) allowed the mural to remain on display. No matter what side you favor in this conflict, it presents a case where an artwork is more than just an artwork, and where a question about preservation animated a community (and many interested onlookers around the nation). How do we reckon with our history and what does it say about who we are? What we choose to save plays a pivotal role in how we answer these questions.

Why are we so frequently confronted by conflicts and questions about saving things? Once you're attentive to conservation issues, you quickly realize they are ubiquitous. Perhaps the reason is that wanting to maintain the things we care about seems to be a central part of caring about them in the first place. As the philosopher Samuel Scheffler puts the point: "It is difficult to understand how human beings could have values at all if they did not have conservative impulses. What would it mean to value things, but in general, to see no reason of any kind to sustain them or retain them or preserve them or extend them into the future?"[3] If I tell you that I cherish something, but take no steps to protect or maintain it, you might not only criticize my laziness or carelessness but reasonably come to question whether I really care about the thing at all. Our lives are defined by what we care about, and so it is no surprise that hanging onto cherished things would be among our paramount concerns.

But while a complete disregard for preserving something can call into question whether we really value it at all, an obsessive preoccupation with conservation can surprisingly present us with the same puzzle. The problem is that the most reliable and effective ways of conserving the things we care about can separate us from the source of their value. Think of a cherished stuffed animal from your childhood (a "lovey" as I learned to call them from my daughter's daycare). You loved that lovey half to death, and it required some maintenance to keep it in shape. Maybe a button eye had to be sewn on again, stuffing re-stuffed, a rip stitched. But would you put that lovey on the high shelf? Would you encase it in glass? To what lengths would you go to arrest deterioration? Certainly, no interventions that would prevent snuggles.

This is the problem that sits at the heart of conservation. It's the need to impede destruction that foists the imperative of preservation upon us, but if we aren't careful, avoiding loss can come to dominate our relationship to threatened things. Animated only by the desire to save them, we can lose sight of why these things mattered to us in the first place.[4] Sure, you can keep your lovey unassailably safe in a hermetically sealed vault, but what's the point?

There is so much that we want to save. Species, buildings, traditions, landscapes, dances, artworks, video games, instruments. We care about a vast array of things, and they are all subject to the threat of loss.[5] This book is about how to navigate that threat, but it isn't a guide to saving anything in particular. You can read expert practical analyses that offer instructions about how to conserve all kinds of specific things, in fields from conservation biology to clothes mending to underwater archaeology, and everything in between. This book, rather, is about how it all hangs together.[6] That's not to say that the book won't have practical implications—I hope that it will! But they are implications that stem from how we think about conservation rather than from technical instructions. The *what to save* of the book's title is not the promise of a list, but an invitation to reflect. And while I aim to be informed by the work of various conservation fields, this book isn't *about* any of those fields: it's about the idea of conservation itself.[7] Likewise, this book isn't *for* conservators, though I hope they'll find it thought-provoking. Rather, it's for anyone who is animated by questions about saving things (which I think is ultimately all of us, even if we don't realize it at first).

When you think about the Conservation Department at an art museum, for example, it's natural to conceive of conservation primarily in terms of technical expertise rather than philosophical questions that anyone might ponder. We want the Monet painting to be conserved and the folks in the Conservation Department have the skills and training to do it well. Simple! But thinking about conservation in these terms can obscure a set of more difficult questions—*normative* questions about how we *should* approach conservation. Normative questions are about how things ought to be rather than about how they actually are. A descriptive question is: do people sometimes break their promises? We can use observation, or social scientific methods if we want to get fancy, to answer that question. Yup, people sometimes break promises. A normative question takes a different tack: is it always *wrong* to break a promise? That's a question that we can't answer through observation alone. It's a question about what we *should* do (or shouldn't!), not about what people actually do, and so there's an ineliminable element of judgment required to answer it, some appeal to values, an understanding of how the world ought to be. That's all it is for a question or statement to be normative, as opposed to just descriptive. Some normative questions (like the one about promising) have an *ethical* dimension to them, concerning not only what we ought to do but whether it would be right or wrong to do it. Not every normative statement is straightforwardly ethical (consider, "if you're hungry, you should eat something"). But even a straightforward piece of practical advice like eat when you're hungry has ethical values lurking under the surface. Why avoid hunger? Why nourish yourself? Follow the paths that these questions point us down and you're bound to hit some moral values (the value of avoiding pain, the meaning of life, etc.) even when they're not apparent at first glance.

We can expand our perspective on the complexity of conservation by posing the following set of normative questions:

What things should be conserved?
Why should they be conserved?
From what changes should they be conserved?
How should they be conserved?
Who should conserve them?
For whose sake should they be conserved?

Let's go back to the example of the Monet painting. Rather than just noting with approval that the Conservation Department does its job if it conserves the painting, we might describe the situation this way: A professional conservator conserves the Monet painting in order to maintain the painting's visual appearance. She does this by cleaning it carefully to prevent the dirtying and obscuring of the pigments so that anyone who visits the museum can view the painting with its original vibrant hues.

Framed this way, we can see that this description assumes a set of answers to the normative questions we posed. A conservator with technical expertise *ought* to perform the task of conservation in this case. Why? Because we ought to conserve great works of art. We do this so that people can experience the aesthetic achievements of exemplary artworks, and they can only do this if they can see the works in as close to their original intended form as possible: successfully maintaining the opportunity for these experiences takes skill and training. Anyone in the world might rightly value such an experience, which is why we think it is good that the work is located in a museum that is open to the public.

While this description assumes a set of answers to our conservation questions (and a common set when it comes to famous artworks), they're not necessarily the correct ones. All of them are up for debate, and changing any one of these variables might in turn alter the others.[8] Imagine if we didn't care whether anyone alive today could view the painting; rather, our goal was for it to be a contribution to a time capsule that would serve as a record of human artistic achievement in late 19th- to early 20th-century visual art. In that case, we might seal the painting in a lightless vault and bury it underground, or even shoot it into outer space in the hope that some creature, at some time, might know this about us. Or consider if we didn't care about maintaining the work's original appearance—we only cared about the continued existence of the physical object, in whatever state, no matter what it looked like. In such a case, the need for expert conservation might in turn fade away. We might be perfectly happy with an amateur conservation effort, such as the infamous "beast Jesus" restored by Cecilia Giménez in Borja, Spain, which bears laughably little resemblance to the original.[9] Giménez's restoration was even lampooned on *Saturday Night Live*, where Kate McKinnon (in the guise of the restorer herself) aptly described it as having a "round monkey face" and "dead black eyes" that "look like a shark." The world was aghast at this "restoration," not only because it was comically bad as an attempt to preserve the original look of the fresco, but because it conflicts with common and deeply held views about what the goals of conservation should be and how we should conceive of success in achieving them. Answer these questions differently and our judgments about the outcome would change. It can be difficult to see this in the beast Jesus case, where the answers to the relevant questions are so entrenched, but if we look to a different instance in which the same kinds of questions arise, it is easier to see the variety of possible answers.

People have starkly divergent views about the conservation of historic buildings. Some think that it is essential to maintain every facet of the building's period character, down to paint colors and furniture selection, whereas others care about maintaining the character of the facade but adopt a radically modern approach to the interior.[10] There are many examples of homes and offices where the building exterior suggests a perfectly preserved period structure, but the interior, or even another face of the building, is all metal and glass and contemporary touches. Some will find this reimagining exciting; others will consider it a vulgar affront to the conservation of historic buildings. Unlike in the beast Jesus example, it's easier in this case to see that there are different ways of answering the normative question about what a successful conservation should aim to do. This illustrates what I take to be one of the virtues of a broad approach to thinking through conservation issues. When we limit our focus to a particular kind of conservation case, it's easier to get trapped in conventional ways of thinking. Comparison across conservation contexts, sometimes even radically different ones, may reveal surprising and exciting, if counterintuitive, possibilities. As the philosopher Jeanette Bicknell asks: "How might our aesthetic practices change if we rejected the dichotomy between stable and ephemeral and strove to appreciate and enjoy even the most solid of structures as if it were to disappear in a short time?"[11] You savor an ice cream cone, the cool combination of salty sweet caramel on your lips and tongue. You revel in devouring the ice cream; or to put it in unusual terms, destroying it. What might it mean to appreciate a painting or a building in that way, or at least something like that way? What *could* it even mean? And how would it influence our approach to conserving the things we care about? We're going to sit with these kinds of questions.

I've used both the terms "conservation" and "preservation" in the discussion so far, and we'll continue to encounter a constellation of related ideas as our investigation unfolds: restoration, repair, replication, reproduction, protection, maintenance, etc.[12] In some fields, the preservation/conservation divide is not just terminological, but ideological, and the terms don't always pick out the same commitments across different disciplines. In some contexts, preservation implies a commitment to arresting any and all change, whereas conservation is more concerned with the management of change.[13] In other contexts, preservation has been understood as being for the sake of the thing to be preserved, whereas conservation is protecting it for the sake of something or someone else (e.g., preserving nature for its own sake vs. conserving it for human enjoyment).[14] As should be clear from the foregoing discussion, I'm interested in exploring all of these questions and the ramifications of how we answer them, so I don't want my use of terms to presuppose an answer to any of them. I will generally use *conservation* as a catch-all term that includes preservation and restoration,[15] but I will also sometimes use *preservation* to refer to whether to save something at all, and *conservation* to refer to how we go about saving it, acknowledging that there is conceptual overlap between these categories. What's ultimately at stake is the common root between conservation and preservation; namely, *keeping* things. What that amounts to in practice will depend on how we answer the six conservation questions.

Each of these six questions (What? Why? From what? How? By whom? For whom?) will be discussed in its own chapter, but they're not neatly separable from one another. What we should preserve obviously has implications for how we should conserve it. Why we should conserve something can inform for whom it should be conserved. Sometimes we'll need to hold the answers to some questions fixed in order to focus on one question at a time. This can be artificial but also illuminating, the same way we might control for variables in a lab experiment. A lab experiment isn't the same as a field experiment, but that doesn't mean there aren't important insights to glean from it. The order of the chapters will also allow our inquiry to build as we proceed—understanding the space of potential answers to the earlier questions will shape how we answer the later ones.

While the idea of conserving something naturally evokes action (*doing* something with respect to a conservation candidate), an essential dimension of our discussion in this book will also concern attitudes (*feeling* something about the conservation candidate). Anything we *care* about might generate an interest in conserving it, and I aim to keep the notion of caring firmly in view throughout this discussion—conservation often evokes strong emotions precisely because it concerns things that we are deeply invested in keeping or saving or holding onto. Once we become alive to the possibility of losing the things we care about, the urge to preserve is a natural, and sometimes startling, impulse. Consider this remarkable 1944 quote from Sir Harold Nicholson:

> I should assuredly be prepared to be shot against a wall if I were certain that by such a sacrifice I could preserve the Giotto frescoes; nor should I hesitate for an instant (were such a decision ever open to me) to save St. Mark's even if I were aware that by so doing I should bring death to my sons . . . My attitude would be governed by a principle which is surely incontrovertible. The irreplaceable is more important than the replaceable, and the loss of even the most valued human life is ultimately less disastrous than the loss of something which in no circumstances can ever be created again.[16]

What's so powerful (and perhaps disturbing) about this passage isn't anything that Nicholson does (the actions he's entertaining are hypothetical). Nicholson is clearly very passionate about conservation! At the same time, there's something unsettling about framing that commitment in a way that gives it priority over the lives of his own children, even if the point is intended as hyperbolic rhetoric. This passage reminds us that the attitudes we adopt toward the prospect of change and loss are also an important dimension of our response to it. Ethics in general is concerned with what we think and feel in addition to what we do.[17] Once we are in this frame of mind, the space of possible responses to the threat of loss blossoms—we have many more options than just whether or not to attempt arresting decay. Whether we're thinking about keepsakes and childhood haunts or fine artworks and ancient artifacts, our relationships with the things we might conserve, in all their messy emotional complexity, require as much careful attention as the things themselves.

This kind of people-centered approach to conservation has been gaining ground in the scholarly literature and in various fields of conservation management. Paper conservator and theorist Salvador Muñoz Viñas describes a shift in the field of heritage conservation from scientific study to values, driven by a desire to prioritize the diverse kinds of significance that heritage has for different people, in contrast with an exclusive focus on "historical truth" or scientific study of an artifact's material properties.[18] He suggests that this pushes conservation in a more sociological direction, since people's values will be subjective, and (presumably) we'll need sociological methods to determine what those values are.

On the one hand, sociological and anthropological investigation into what people value and why is essential: we should welcome empirical inquiry that helps us understand the shape and variety of human values. On the other hand, the recognition of the role that values play in contemporary conservation should just as importantly herald a key role for philosophical inquiry alongside social-scientific investigation. Because there's a difference between descriptive investigation of what people's value *are*, and normative exploration of what people's values *ought to be*, empirical inquiry on its own will be insufficient to address the latter set of issues. Philosophy as a discipline is well-positioned to explore the nature and structure of these values, independently from determining their content. Muñoz Viñas assumes that values are subjective, adopting a simplistic binary between scientific objectivity and subjective evaluation. But things are not so simple. To be sure, some evaluations are purely subjective. If I tell you that this ice cream tastes sweet to me, I can't be wrong about that. The truth of the claim is relative to facts about me, the subject, who has the experience: that the ice cream tastes sweet *to me* is a subjective truth. While it's possible that I could misuse the language (maybe the ice cream tastes rancid, and I say it tastes "sweet" because I haven't mastered that term in English), *how* it actually tastes depends on me. If I tell you that the ice cream tastes sweet, and I'm using the language properly, it would be absurd for you to tell me that I'm wrong. Imagine you getting up in my face and insisting that the ice cream doesn't taste sweet to me! I'd assume you were joking or had lost your grip. How would you know how it tastes to me?

But even though the truth of our subjective experience is, appropriately, subjective, many of our important values extend beyond the boundaries of our subjective experiences. They are shared, and we *want* them to be shared. Imagine that we go to the movies together: I think the movie is hysterical, but you find it juvenile. We could just leave things at that, but if we care about movies (and each other), voicing our subjective experiences can be the *start* of a conversation that goes beyond those initial reports and aspires to a *mutual* understanding of the movie and its humor.[19] Through discussing, critiquing, contextualizing, comparing, we might come to revise our initial judgments about the movie. Maybe we'll settle on a judgment that we agree on; maybe we won't. Perhaps we'll come to better appreciate each other's point of view. The point is that we can *seek* a kind of *intersubjective* truth, one that's shared between us, and philosophy can help orient us to that tricky task (though never simply solve it for us).[20] This isn't the same kind of inquiry as scientific investigation of mind-independent facts, truths that would be "objective" in the sense that they would be true independently of any facts about our subjective experience—but nor is it *trying* to be objective in this way. Normative inquiry is often messier than objective inquiry, and requires different kinds of methods, but that doesn't make it merely subjective. And crucially, we don't tend to treat it as if it were merely subjective. Questions above significance and value, whether historical, aesthetic, or ethical, often generate the disputes that we care about the most. If you thought that the normative claim "killing people for fun is wrong" was merely subjectively true, much of social life would have to be a complete mystery to you.

So, yes, conservation is largely about values, and that makes questions about it challenging. But it doesn't mean the answers will be merely subjective, and careful, creative thinking can help to guide our discussions about those answers. If you're game for that task, then the chapters ahead are for you. What should we conserve? Why? From what? How? By whom? For whom? I can't offer definitive answers to any of these questions—they are as much yours to consider as mine. Rather, I hope that by examining how threads from diverse contexts and intellectual traditions hang together, we might discover patterns that can inform our conservation thinking. To anticipate, the patterns I see emerging look like this: The need for conservation is impelled by the threat of loss, and the desire to prevent change stems from the hope of preserving our sense of self. But what really matters is not repelling any alteration, but confronting change on terms that we can accept—terms that grant us a role in managing the forces of change instead of being dominated by them. It is through becoming part of the process of change that conservation empowers us to mold the future of things we care about, and in so doing, shape who we will be.

These patterns allow ample room for variation, but they also offer a vision that constrains what conservation should look like. They acknowledge that there is much about which we can still disagree, but perhaps, also, some truths that we can share.

2 | FROM HEIRLOOMS TO TACOS

WHAT SHOULD WE SAVE?

Picking over heirlooms, mementos, and knick-knacks, the Joad family in John Steinbeck's *The Grapes of Wrath* must decide what they can save. Forced to flee their home by the pressures of environmental and economic hardship, they have little room left for the sentimental, and it weighs on them:

> The women sat among the doomed things, turning them over and looking past them and back. This book. My father had it. He liked a book.
>
> Pilgrim's Progress. Used to read it. Got his name in it. And his pipe—still smells rank. And this picture—an angel. I looked at that before the fust [*sic*] three come—didn't seem to do much good. Think we could get this china dog in? Aunt Sadie brought it from the St. Louis Fair. See? Wrote right on it. No, I guess not. Here's a letter my brother wrote the day before he died. Here's an old-time hat. These feathers—never got to use them. No, there isn't room.
>
> How can we live without our lives? How will we know it's us without our past?[1]

As this passage makes manifest, questions about what to save are complex and vexed, even when they only concern personal possessions. We face practical constraints, for one: we don't have space to save everything. But we also face questions about what exactly it is that we're trying to hang onto. You'd think there would be a substantial difference between wanting to save some old stuff and wanting to save your sense of self! How will we know who we are without this old hat? Put that way, the question sounds ridiculous. And yet those two conservation candidates are united in the Steinbeck passage, and the linkage between them is clearly not intended to be absurd. Many of us feel that the material collections of our lives and the lives that preceded ours have an emotional residue that helps to remind us of our relationships and commitments, buoy us during challenging episodes, and anchor us in place and time. The Joads have got it right—all conservation questions are ultimately bound up with the prospect of saving ourselves, whether our perceived connections to the past or our aspirations about the kind of people we aim to be. But whether actually conserving *things* is always the best route to achieving this goal is decidedly less clear.

It takes some philosophical sleuthing to arrive at this conclusion. The conundrum confronted by the Joads is emblematic of conservation problems faced more broadly, whether in museums and libraries or historic districts and parklands. It commences with the question of what we should save in the first place. While I'll show that a concern with identity tends to lurk in the background of this question, it's not the focus that is typically foregrounded in talk about conservation. To recognize the role played by identity, we'll first need to consider some more familiar targets of conservation: material, memory, history, authenticity.

—

The natural starting point for answering the question about what to conserve is the obvious candidate: the things that we should conserve are the things that we care about keeping. The emphasis here is on the thing itself, whatever it may be, and that's a standard way of thinking about conservation candidates in a wide range of material contexts (saving intangible things such as practices and traditions complicates the picture further, as we'll see). We care about paintings and buildings and cars and pottery and books and any number of other physical things, and when we consider conserving these things, it seems what we want to do is prevent these particular material objects from degrading or breaking or disappearing.

But when we press on what it is about these material objects that we want to save, the appearance of an obvious answer vanishes. Do we really want to save the physical object itself, the original component materials? Or do we want to conserve how the object looked at a particular time, or a specific function that it had, even if that means replacing parts or modifying the object to maintain a certain appearance or operability? Is what matters to us about these things really what we can perceive about them (their appearance, texture, sound, smell, taste) or is it how they make us feel, what they help us remember? Are we ultimately interested in conserving an object's meaning (to us? to anyone?), and might that be a different goal from conserving the thing itself?

We can begin to reveal the tensions among different ways of answering these questions by considering The Ship of Theseus, a story recounted by Plutarch. Imagine that Theseus sails the Mediterranean in a wooden ship, and every time a plank grows weak or rotten, he replaces it. Eventually, all of the planks have been replaced, and the ship Theseus sails is composed of entirely new material. We can also imagine that the discarded planks are meanwhile reassembled into another ship, somewhat worse for wear, but otherwise identical to the ship that Theseus is currently sailing, though these planks were only involved in his previous voyages.[2]

This case is an old chestnut in philosophy, used for probing paradoxes surrounding the identity of objects that change over time. From a conservation perspective, we don't (yet) need to get mired in questions about which ship is "the real" ship of Theseus. Rather, we can use this case to get clarity on what we're interested in conserving. Say that Theseus only has the resources to conserve one ship (let's call them the seaworthy ship and the relic). If Theseus cares about the activity and experience of sailing, then he will focus his attention on the seaworthy ship. If he cares about the memories and associations of the original material, he will conserve the relic. Of course, he might well care about both, and the constraints of limited resources will force him to choose. But he might also pause at this point and wonder: even though he is sentimental about the relic, and would conserve it given the resources, does he need to retain the whole original ship in order to maintain an emotional connection with his previous voyages? Might preserving one oar be enough? He could mount it above a doorway (maybe next to the Minotaur's head). Returning to

Steinbeck, could the Joads keep only the old hat without giving up the connection to their past and identity?

The Joads, as it turns out, don't take *anything* from their pile of mementos, but their response to the problem is poetic and revealing. "'No,' They say. 'Leave it. Burn it.' They sat and looked at it and burned it into their memories."[3] After questioning whether they can retain their identity without their material links to the past, they decide that they can, even if it's not the outcome they'd prefer. It is an act of will to remember, and the Joads choose memory. This is evocative of an influential claim made by heritage scholar Laurajane Smith, writing about the function of heritage sites: "Heritage . . . is a cultural process that engages with acts of remembering that work to create ways to understand and engage with the present, and the sites themselves are cultural tools that can facilitate, but are not necessarily vital for, this process."[4] If keepsakes and heirlooms aren't essential for acts of remembering, and memory is what we care about, then maybe we can sometimes get on without old stuff. Perhaps for the Joads the link between object and memory will be enhanced by commemorating the loss through conflagration. The point isn't to disparage our connection to physical mementos themselves, or to encourage us not to regret their loss—we shouldn't be eager to surrender anything to the flames. But it is to establish that when the chips are down, we have the ability to hold on to some of what makes sentimental objects worth caring about in the first place. A prop for memory isn't the same as the memory itself—mental connections can survive the loss of their physical correlates. Seeing this, as the Joads do, is essential for moving past an undue obsession with loss, especially in the many cases in which loss is unavoidable. How do we acknowledge, document, mourn, or even celebrate loss rather than just trying to avoid it? These are questions we'll continue to rub up against until their surfaces are worn. But as with antiques, those patinas can be valuable—hard questions merit revisiting again and again.

Vehicles for memory are only one means of conveyance to recollection. When the choice we face isn't quite so stark as the one confronted by the Joads, it is especially helpful to keep this in mind. Many of us are familiar with the experience of being left *heaps* of stuff by our parents, grandparents, or other extended family. It's not unusual to feel guilty about the prospect of failing to conserve these items, and sometimes it's possible to just hang onto the lot of it. We have boxes gathering dust in attics or basements because we can't bear the thought of discarding things that our forebears cared about. But if what we want to save is really the connection to our loved ones, then we need to ask how that connection depends on the things they leave behind. Does keeping our loved ones in our hearts and minds require keeping all of their stuff?

One way to approach these questions is by thinking through what makes particular items *distinctive.* I don't just mean different in type or appearance or quantity—there's a boring sense in which the two identical pencils on my desk are distinct from one another. Rather, we need to think about what makes things have distinctive *meanings* for us. Certain objects can have strong connections with particular memories, whereas others may not cue up any associations at all. If you already have a cherished set of family mementos, discovering yet another box of unfamiliar knickknacks can be frustrating—you might have a felt sense of obligation to preserve your family's things, but the force of that duty can chafe against the lack of personal significance that those items hold for you. Prioritizing things that ring a bell, though, can at least offer a principle for separating the wheat from the chaff if circumstances require it. To riff on Marie Kondo, we might ask of any heirloom: "Does it spark recognition?" You don't risk sacrificing your memories if the objects in question are unrecognizable—the threat of loss only rears its head when there's something to lose.

—

Caring about the mental associations borne by physical objects is different, though, from caring for the history of the object itself. Even if you don't have any associations with an heirloom, just knowing that it's part of your family history might keep you from consigning it to the dustbin. In this case, an exclusive focus on conserving objects as prompts for memory would be missing something. The philosopher G. A. Cohen captures this kind of concern with the very history of an object, describing a cherished old eraser, of all things. He writes:

> I would hate to lose this eraser. I would hate that even if I knew that it could be readily replaced, not only, if I so wished, by a pristine cubical one, but even by one of precisely the same off-round shape and the same dingy colour that my eraser has now acquired. There is no feature that stands apart from its history that makes me want to keep this eraser. I want my eraser, with its history. What could be more human than that?[5]

The idea that objects with a certain history warrant conservation is a powerful instance of the purported link between human nature and cherishing the past. People love sci-fi epics about time travel, imagining the possibility of hopping back and forth between disparate times. But our interest in fantastical time travel stories can distract from the more everyday but no less remarkable fact that old objects are also time travelers. Yes, they only move forward in time at the same pace we do, but because they started that journey before we did (sometimes long before), they connect us to times that we were never present for and can never visit.[6] And while it's not the world of sci-fi or fantasy, there's something close to magical about it.

That doesn't mean our interest in the history of objects needs to involve "magical thinking" though. When we're in the grip of magical thinking, we attribute properties to objects that they don't actually possess. We might apply a principle of "contagion" to an object, as if a person's essence somehow rubs off on the things that they've held.[7] Think about the rock star throwing his sweaty headband into the crowd of adoring fans, his devotees grasping for it as if he had perspired his soul into the fabric. This phenomenon is only accentuated when we move beyond contemporary associations (whether personal or popular) to artifacts connected to significant historical events or to times almost unimaginably removed from our own—Abraham Lincoln's pen, or an ancient fossil.[8] But we don't need to believe in any mystical transference of essences to understand the wonder that these objects can produce.

The philosopher Carolyn Korsmeyer describes ancient ruins as "huge and immobile witnesses to the past."[9] While the idea of an ancient structure as a witness flirts with metaphors of vision (objects can't literally witness anything), there is nothing metaphorical about the fact that the structure was *there*, in the past. Though it lacks the eyes to see, it nevertheless bears witness to the passage of time through the evidence of its own degradation. Scientists discuss the parallel notion of a *witness tree*, a tree whose age allows us to learn about the history of a place, "reporting" the dynamics of environmental conditions through their accumulated effects.[10] Korsmeyer argues that the "brute materiality" of ruins is in turn key to our appreciation of them. The fact that they were present in the past carries over to the wonder of our own opportunity to be present among them now. She writes: "the ruin itself engages all the senses; it can be touched, moved around, even climbed on . . . the intense awareness of *being there*, within touching distance, deepens or perhaps even provides the thrill of a ruin."[11]

This evocative description requires that we consider two different ways we might value an object for its history. The desirable qualities we appreciate in the visceral ways that material things show the signs of age—wear and tear, deterioration and decay—can be called *age value*. *Historical value*, in contrast, captures the sense in which objects are representative of and involved with past moments of particular historical significance.[12] An object can clearly have both age value and historical value, but the two also come apart. A ruin might have significant age value, but little historical importance; a carefully conserved painting or artifact might have substantial historical significance while showing little sign of being weathered by age.

But do we actually need the original material to secure either historical value or age value? It's common for commentators to disparage replicas in the context of these values. The Institute for Digital Archaeology, housed at Oxford University, created a scaled 3D printed replica of the triumphal arch at Ancient Palmyra after it was destroyed by the terrorist group Daesh (also known as ISIS). The arch was displayed at Trafalgar Square in London with the support of Syria's director of antiquities, in theory as a bid to raise awareness, and with an eye toward future restoration.[13] The project was panned from many corners, though, portrayed as a subpar stand-in that contributes to the "Disneyfication" of history.[14] The printed Palmyra arch can seem to function as a kind of *replacement* for the original on the heels of its destruction, and I suspect this is often what generates animosity toward replicas even when they are only being considered in the hypothetical. To say "I would never accept a replica of this cherished item, even if it looked exactly the same!" implicitly evokes the idea that the role of a replica is to replace, and that it would be hopeless in capturing the aura of the original.[15] But we can grant that a copy lacks the exact same value as an original without concluding that it is a worthless knockoff. Much depends on what we want a replica to *do*, and there are many options besides replacing originals. We'll look more closely at the work that replicas can do in Chapter 4; for now, our focus is on what replicas can teach us about *what* we should save. Depending on what we're after, a replica may be as good as or better than the real deal.

We can see this by turning back to Korsmeyer, who is wary of faux aging. She writes: "While marks of age may be replicated—chips, cracks, signs of wear—they do not have the same effect as something that is truly aged."[16] This is because they don't actually embody the history that they appear to, and Korsmeyer proposes a kind of test for assessing this value. If we revealed to an eager throng of onlookers that the object they were peering at (say, the Mona Lisa) was a replica, would the wonder they feel dissipate? If so, it suggests there's something important about the genuine article, the object with its actual history, that is lacked by a stand-in that merely looks the same. There's something missing from replicas, the conjecture goes, even if they're indistinguishable from originals, and the lack of interest that audiences evince for copies is indicative of that absence.

There's something initially puzzling about Korsmeyer's test, though. She asserts that faux age doesn't have the same effect as something truly aged, but requires that the audience *know* the object isn't truly aged in order to see this. If replicated age produces a sense of wonder in its audience, why is that not enough? If the spell is only broken by revealing the otherwise indiscernible nature of the replicas, does that actually mean that replicas lack the same effect as originals? Archaeologist Cornelius Holtof writes: "visitors to archaeological sites or museums experience authenticity and aura in front of originals to exactly the same degree as they do in front of very good reproductions or copies—as long as they do not know them to be reproductions or copies."[17] The wonder of magic tricks can also be spoiled by revealing how their artifice operates—that doesn't devalue the original trick.

The limitation of this analogy is that we approach magic tricks with the implicit understanding that we are being fooled in some way, and we know that the pleasure of the performance depends on maintaining the secrecy of the trick—we don't really *want* to know how it's done. In this sense, magic wears its deceitfulness on its sleeve, and we're in on the scheme. But when we approach an artifact believing that it has a particular history, there is no analogous acknowledgment that there is trickery afoot. A museum that serves up a copy masquerading as an original betrays our trust. This can in turn undermine the effect of being in the replica's presence, even if we may first be fooled into beholding it in wonder. What matters, then, isn't so much anything about the replica itself, but rather, how it is presented to us, and how this influences our expectations. Consider by comparison the case of the 20th-century artist Han van Meegeren, who successfully passed off his own original paintings as the work of Baroque masters such as de Hooch, Hals, and most infamously, Vermeer.[18] The forgeries were so convincing that van Meegeren's deceit was only revealed when he was brought to trial after World War II for selling "Dutch national treasures" to the Nazis. Van Meegeren explained that he hadn't actually committed this crime, because the supposed Dutch national treasures were in fact his own modern forgeries! Though he had a defense against this particular charge, there are of course other worries we might have about the forgeries. The problem can't be their lack of originality, though, because they were original paintings by van Meegeren, just deceitfully misattributed. Rather, what's objectionable about the whole affair is that van Meegeren's paintings misrepresent the nature of his achievement. For van Meegeren to create works in the 20th century that are evocative of Baroque paintings is a different kind of achievement than to create those paintings in the 17th century, and this is what he misrepresents when he falsely attributes his work to an earlier period.[19]

There is a similar sense in which fabricated signs of age mislead, though mistaken achievement isn't the proper category for assessing the deception. If we approach the look of age with the expectation that it has been wrought by the long passage of time, only to learn that it's the result of a clever chemical treatment from a few years back, then we might still say it is a kind of misrepresentation—we are invited to view the object as old, when it is in fact a modern facsimile.

What becomes clear in considering Korsmeyer's test, though, is that if we lack the particular expectations about the object that she describes, then we in turn can't assume that viewers will be disappointed by replicas, or unable to discover a sense of wonder in them.[20] Views about the significance of originals have varied with time and place. Plaster casts, for instance, have a long history of being collected and displayed in Western museums and private collections. There are a variety of concepts used for copies in China and Japan that do not carry any pejorative connotation.[21] Twenty-twenty-two marked the 100th anniversary of the discovery of the tomb of Tutankhamun: touring exhibitions of "Tut's treasures" consisting of nothing but replicas drew millions of visitors.[22] It may be true that wonder dissipates when expectations are disappointed, but this doesn't tell us anything about whether we ought to seek out and expect originals. Consider that if I visit the studio of a famous art forger to view her latest work, and I stare in awe at her masterful forgery of a Raphael, I will also be disappointed to learn that the painting I'm looking at is an authentic Raphael after all, and not the product of the forger's genius. These cases indicate that there can be something magical about the genuine article, but not that there must be. It depends on what we're looking for.

—

We have finally arrived at that unruly concept—authenticity—that haunts discussions of conservation, and which will be present throughout all the chapters of this book, sometimes explicitly, sometimes as a silent specter. Authenticity is so often the traveling companion of conservation in part because a concern with either only arises when we are faced with the threat of their loss.[23] When we encounter the prospect of destruction or disappearance, what things should we aim to preserve? Well, the authentic ones. Naturally.

Some commentators insist that talk of authenticity is primarily a distraction. According to what he calls the "tautological argument," Salvador Muñoz Viñas asserts that everything is in fact authentic, and as such there is little to be gained by attending to authenticity when it comes to conservation issues. The argument goes like this. To say that something is "inauthentic" is to say that it is false. But "if something exists, it is real, it is authentic, it *cannot* be false."[24] Thus no conservation effort can make something any more or less authentic: the authentic state of the object is just the state that it is in. This argument shares the same spirit as worries about environmental restoration "faking nature," though the polarity is reversed.[25] The idea is that if we define nature as the absence of humanity (think of the idea of "pristine, untouched wilderness"), then no human effort to restore nature can be successful: by definition, human intervention is inherently unnatural. If we try to restore nature, we just further its descent into the unnatural, polluting it with our nonnatural humanness.

As many have pointed out however, humans are *part* of nature, so it's bizarre to define us right out of the natural world. It's only that definition that renders us unable to aid in the process of nature's restoration.[26] The idea that humans are separate from nature has had particularly adverse effects for Indigenous communities in the context of nature conservation, leading to displacement from sites that are today considered emblematic of successfully "saving nature," such as Yellowstone and Yosemite National Parks.[27] We'll return to the relationship between conservation and marginalized communities in later chapters. For now, note that the mistake that is made in defining nature in opposition to humanity is analogous to that made by defining authenticity as a thing simply "being what it is." They are both "tautological arguments," because the consequence is a direct function of how nature or authenticity is defined. But like the definition of nature mooted earlier, Muñoz Viñas's definition of authenticity sorely mistakes the role that a concern with authenticity plays in our lives. When we ask after the authenticity of a thing, we're not interested in the tautological answer that it is the thing that it is, but whether it is a certain *kind* of thing, one that we may have a particular interest in. We ask "is this authentic?" with an understanding that the answer could be "no."

Consider this point in the context of the search for authentic foods. According to one recent account of culinary authenticity, what it means for a dish to be authentic is for it to pass as an instance of a certain culinary style in the relevant context (along with meeting some basic provenance requirements about how it was produced and the ingredients used).[28] In other words, to ask whether a dish is authentic is to ask whether it properly belongs in a certain culinary category. To ask whether the plate of Viet-Cajun barbecue before me is authentic is to ask whether it would pass as Viet-Cajun style barbecue in the relevant cultural context (say, the Viet-Cajun barbecue scene in Houston, Texas).

Importantly, authenticity on this account is a purely descriptive concept. Whether or not authenticity is *valuable* depends entirely on whether the membership category we're considering is of interest to us. If I'm on a quest to sample Viet-Cajun barbecue specifically, then I'll regard it as a good thing that this particular dish is authentic. If I'm looking for East Texas style hickory smoked barbecue, then not so much.

One virtue of this understanding of authenticity is that it offers a framework for seeing how expectations about authenticity operate normatively. Even if it's merely a descriptive question whether an object correctly counts as belonging to a particular category (there's nothing inherently good about it), we can see the way that our expectations about category membership can come to matter. The writer John Paul Brammer recounts the story of the unique Kansas City taco, and how it is being stamped out by consumer demand for "authentic" Mexican tacos. He concludes: " 'authenticity' is restrictive. It limits the imaginations of nonwhite people."[29] A caricature of an "authentic" taco fixed in peoples' minds makes it harder to recognize variations and innovations that emerge from diasporic food traditions, and this expectation can then become imposed on communities in a constraining way.

Brammer's point here is important, but the problem isn't with the concept of authenticity, as he suggests. The problem is with the imposition of a descriptive category on an item that clearly doesn't fit the description. What Brammer is objecting to is the way that interest in Mexican-style tacos is erasing Kansas City–style tacos, and the idea of authenticity could as easily be marshaled in defense of the Kansas City–style taco as against it. Save authentic Kansas City–style tacos! Authenticity isn't the enemy in this story—rather, the problem is the way (and to what) the concept is being applied. But analyzing that situation is not aided by simply claiming that both kinds of tacos are authentically what they are.

As we first saw in the discussion of replicas and faux age, "inauthenticity" can involve not only a failure to belong to a category, but the *misrepresentation* of a thing belonging to a category when it actually doesn't. If we present a Mexican-style taco as a Kansas City–style taco, we've done something misleading. We can help diagnose the mistake in Muñoz Viñas's tautological argument by acknowledging a distinction that often arises in the literature about authentic people, rather than authentic objects. When we ask whether a person is being their authentic self, we can distinguish between two different questions: we can ask whether they are being true to themselves ("authenticity"), but we can also ask whether they are accurately representing themselves to others ("sincerity").[30] Muñoz Viñas's tautological argument is focused only on the first question. He contends that any object is being "true to itself," and is thus authentic, because it is the thing that it is. But the question that we're typically interested in when we worry about misrepresentation is the second question—is the object accurately representing its nature to us? Talk of "sincerity" is strained when we shift from people to objects, but the core idea is the same.[31] Viewed through this lens, we can see how some of the cases we considered earlier are clearly bound up with the sincerity side of the authenticity coin. Is this building actually old? Or are the apparent signs of weathering just the result of a recent acid treatment? While it might not make sense to ask if the building is being "sincere," we can ask that question of the people who present the building to us, inviting us to develop expectations about its age.

This is where the promised link between conservation and identity really starts to come into focus. The idea that there is a connection between the authenticity of things and the authenticity of selves crops up across diverse conservation literatures. On one way of thinking about the connection, they are co-constitutive—what we regard as authentic things shapes what it means to be an authentic self, and vice versa.[32] This idea is familiar from the way that people often think about identification with certain subcultures. A *real* punk is committed to an ideal of transgression that generates preferences for certain kinds of music, clothes, and hairstyles. If you simply "dress up" in stuff that you think looks punk, well, you're just a poser. An authentic identity (of whatever kind) is negotiated through caring about and engaging with the correspondingly authentic things, and what makes those things authentic is in part that the relevant people care about them—there are no punk things without punks.[33] So, you can't just construct an authentic identity by, say, buying the right stuff—rather, the process of thinking through and navigating what makes an object or practice count as authentic is itself part of what that identity consists in.

The fraught connection between authentic things and authentic selves is ripe for objectionable forms of group boundary-policing.[34] This is a way in which authenticity talk can be abused, often by individuals with positions of power or privilege within a particular group identity context. The "poser" label, for example, can become a weapon to exclude certain people from the punk scene (which itself seems pretty un-punk if you ask me, but I'm a square, so don't take my word for it).[35]

But it doesn't need to be this way. The idea of authentic things and authentic identities co-constituting each other can offer resources for exploration, experimentation, and discovery as we navigate how to understand our identities. Writing on a similar theme, philosopher Paul C. Taylor discusses W. E. B. Du Bois's reflections on an African song that he learned from his great-grandmother, though Du Bois knew nothing about the song's origins or the meaning of the words. The point Taylor makes is that Du Bois does not use the song as a simple prop to connect him with his African heritage, as if finding connections to authentic African things just creates an authentic African identity. Rather, as Taylor describes it, cultivating a relationship with the song "is less about digging up authentic roots than about self-consciously and creatively assembling resources for inhabiting the present and future."[36] According to this perspective, authenticity is less about the features of things (or people), and more about how we use things to understand our place in the world. They are the materials out of which we build an identity, and in so doing we shape the meaning of those very things.

The common thread running through these different attempts to nail down the meaning of authenticity is that what we regard as authentic, or what we devote our time and attention to conserving (even independently of any discussion of authenticity) is shaped by and shapes in turn a conception of the person or group who makes conservation decisions. Think back to the worry that attempts to restore or rehabilitate nature will simply result in a fake and artificial nature-knockoff. Even if we agreed with the idea that we can't *recreate* nature through restoration, our efforts to try and do right by nature, to make environmentally conscientious decisions, can shape our *relationship to* nature. For example, we can aim to end practices that involve *dominating* nature (such as clear-cutting forests or treating the ocean like a garbage dump), and thus shape our relationship with nature moving forward.[37] We can extend this idea of environmental restoration work as a way of cultivating a relationship with nature to a broader set of conservation and restoration candidates. How and whether we adopt attitudes and practices of care and concern for heirlooms, historical artifacts, or traditions can shape our relationship to the past or to our culture.[38] And just as our relationships with other people constitute a fundamental aspect of our identity, so do our relationships with all of the things we care about. Conservation thus has a role to play in shaping these relationships, with consequences for our understanding of who we are.

Thinking about our relationship with conservation candidates can helpfully foreground the ways that conservation is often an intimate practice, involving deeply felt connections to objects, traditions, and places. The way we *feel* about the things we care about is part of our identity, too, and we can come to see these things as imbued with feelings themselves, a kind of meeting of the emotions. As we've seen, there doesn't need to be anything magical or mystical about the idea of an aura—a further way to think about that notion is in terms of how it evokes the way things can feel to us, a sense that extends beyond just artworks, artifacts, or other objects. Places, for example, can have a certain *character*, and when it comes to preserving them, this is often what we're trying to save. As architect and preservationist Daniel Bluestone puts it, historic preservation sites "have in common a palpable, tangible, physical character that is valued by preservationists and devalued by their opponents."[39] According to Bluestone, this character is both the focus of preservationists' attention as well as the means through which they connect sites with the past. "Historic preservation engages history through the palpable character of place."[40]

While "character" is a common term for describing the feel of a place, other concepts in the same vicinity include "atmosphere," "mood," "ambiance," and "vibe." Focusing on the *feel* of places and things can help expand our thinking about what exactly we're interested in conserving, and indicate another avenue through which identity influences the answer. If what we care about is maintaining the character of a place, for instance, then it remains an open question what the relationship might be between conserving that character and saving any of the particular physical components of that place. Perhaps character depends on conserving the material constituents of a place, but maybe that relationship is looser than we might initially assume. Could Theseus's seaworthy ship have the same character as the relic? And if not, how do we understand the point in the replacement of planks at which it lost that character? Maybe there is something special about the historical properties of a thing, but that need not rule out exploring how character or aura could also be secured in other ways.

Consider how we might understand the relationship between character and component parts, between feel and physicality. Writing about the notion of aesthetic character (in contrast with moral character), the philosopher Emily Brady describes it as "an emergent quality from constituent aesthetic qualities, the overall quality that gives a landscape, artwork, or person a distinctive look or feel."[41] This theoretical picture of aesthetic character is compelling, and it helps explain why the character of a place appears particularly susceptible to change at the descriptive level. If your favorite coffee shop replaces their well-loved eclectic couches with sleek, uniform Scandinavian chairs, that may in turn make the place feel less pleasantly grungy, in turn making its character less inviting or comfortable.[42] Relatively small changes in physical composition might bubble up to the character level with substantial consequences for how a place feels. This observation helps explain why people who are attached to certain places will often vehemently oppose *any* alteration to the physical features of the space. While it can seem obsessive at first glance, we can tell a story about how even minor aspects of a space's physical nature can contribute significantly to the place's felt character.

That being said, this focus on aesthetic character doesn't seem to capture the full range of what we might have in mind when we talk about how a place feels. For example, the character of a place might depend in part on who frequents it and how often, which can have aesthetic consequences, but other ramifications as well. A cafe that used to be a local haunt but is suddenly slammed by visitors on a pilgrimage for the donuts they saw on Instagram might feel like it has lost its character for the locals, the distinctive way it used to feel to be there.

These observations indicate that the feel or character of a place, like the aura of authenticity, is relative to audience expectation. But as we have seen, it would be a mistake to think that recognizing the role that perspective plays in the value of character or aura sends us sliding into the realm of pure subjectivity. We can engage with each other about our values and concerns, share perspectives and considerations that might shape each other's views, deliberate together about what expectations are *worth* having. Should we expect that the character of a place will stay the same over time?[43] What is essential to preserve? Here is where we see identity shaping questions about what to save. The answer depends on who we are and who we want to be.

Considering the character of a place helps to illustrate the stakes in answering these questions. Although we might be inclined to think about a place's character in aesthetic terms, aesthetic character and moral character are often woven together. Descriptions such as "seedy" or "hectic" conjure aesthetic images but also invite moral judgments about the safety, cleanliness, or pace of a place.[44] To acknowledge that these character descriptions are relative to different sets of expectations puts us in a position to see that controversies over conserving the character of a place are also disagreements about whose expectations to affirm. Put more bluntly, to preserve the character of a certain place is to take a stand on who belongs in that space.

The significance of this point is ratcheted up when we recognize that many places that are the focus of conservation efforts are public spaces. When it comes to a local coffee shop, the owner has priority in determining how the character of the space will be maintained or altered. But public spaces are different. We can think of space not as inherently public or private, but always subject to competing power dynamics. Privacy concerns the power of exclusion, whereas publicity concerns the power of access.[45] A space that is officially designated as "public" can still have a private character if it is constructed so as to tacitly or explicitly exclude certain populations. For example, consider how maintaining the character of a space through the cultivation of signs and symbols of "Confederate heritage" could serve to exclude Black citizens from an otherwise "public" space.[46] In cases such as this, we can see the explicit way in which the choice to maintain a certain place character is directly related to a community's moral and political commitments, and speaks to the role of those commitments in the community's identity, in determining what they ought to conserve. For example, if the community is committed to racial equality, then conserving symbols of the Confederacy might reasonably be conceived as flatly inconsistent with their professed identity, whatever other reasons there might be in favor of conservation. Perhaps the community might find a way to reconcile their political commitments with conserving aspects of their Confederate history, for example, through the construction of counter-monuments or contextualization.[47] The point for now is not to answer the other conservation questions that this case invites, but to indicate that an answer is firmly couched in an understanding of that community's identity, both now and in the future. That identity will, in part, both determine and be determined by whether they conserve Confederate monuments, why and how they do so, and who is involved in that decision.

Even bracketing the concerns about access and exclusion, cultivating a certain place character can shape norms and expectations about the attitudes that it is appropriate to adopt toward a particular place. Consider some of the fraught ways that "haunted tourism" has entered the historic preservation space—property owners organizing ghost tours or other forms of paranormal sensationalism to draw a crowd. Historic preservation is expensive, especially when you're dealing with buildings and landscapes. Tourism thus has a key role to play in funding historic preservation of the built environment. When visitor numbers flag, ghost hunters may fill the gap left by dwindling architecture buffs. The problem is that cultivating a creepy character for an historic site is often easiest at places that have witnessed significant trauma, such as penitentiaries and sanatoriums. To wreathe these places in an atmosphere of paranormal intrigue invites visitors to view the events that have transpired there as pleasurable spectacle, a potentially perverse attitude to adopt toward victims of institutional injustice. As reporter Elizabeth Yuko aptly asks, is there a way for haunted tourism to honor these past people rather than exploiting them?[48]

How to honor the past rather than exploiting it is a tricky problem that we'll return to in later chapters. For now, the point that I want to emphasize is that how we confront this challenge will shape our identity through the way it molds our attitudes toward the past. We need to attend to how we feel about the past and how we cultivate those feelings in a place through its character. How do we relate to our history (and whose history is it?)? We cannot simply decide that we want to be a certain kind of community and then expect a straightforward procedure that will tell us what and how to conserve. Rather, when we see that our approach to conservation will shape our identity, that identity offers an organizing framework for how the answers to conservation questions hang together. Do we want to worship at the altar of historical artifacts that we conserve immaculately but never see, sealed in vaults for their future protection? Do we want to disregard the material accumulation of our past and embrace an ideal of self-made individualism, unmoored from custom or tradition? These are caricaturish extremes, but they help illustrate how our answers to the core questions of conservation will shape the kinds of attitudes and actions that constitute our identities, both individually and collectively.[49] What we choose to save is about the things we conserve, but it's also about how conserving those things (as opposed to others) shapes our cares and concerns, and thus contributes to the story of who we are.

—

The intertwined relationship between conservation and group identity is especially significant when what's at stake is the preservation of a group itself. The philosopher and Indigenous studies scholar Kyle Powys Whyte outlines the idea of *collective continuance*, "a society's capacity to self-determine how to adapt to change in ways that avoid reasonably preventable harms."[50] Though Whyte's notion of collective continuance is grounded in specific Anishinaabe traditions surrounding the concepts of "interdependence, systems of responsibilities, and migration," he describes collective continuance as "an important value for any society."[51] One of the main ways that cultural groups are formed and maintained is through various group practices and actions, roles and responsibilities. To face the loss of these traditions and responsibilities is by that stroke to threaten the loss of group identity.[52] What these systems of responsibilities are will vary from group to group, but they will often involve care and concerns for specific objects, practices, and places. Whyte paints a picture in which the conservation of particular plants and animals is often essential to the maintenance of practices and traditions that are constitutive of group identity and group flourishing for Indigenous peoples, and other objects, customs, or places can likewise be essential to the preservation of cultural identity for a range of individuals and groups.[53] While collective continuance describes a group's capacity to adapt to change and disruption, only so much loss is consistent with the persistence of group identity. The concept of collective continuance thus affords an excellent model for the way that identity and conservation are co-constitutive—specific conservation candidates play a key role in group practices, but some of those practices in turn involve taking responsibility for the maintenance of conservation candidates. If we are defined in part by our cares and concerns, roles and responsibilities, then conservation is far more than a niche technical interest. What we choose to maintain is part of our ethical core. These consequences for identity maintenance and formation should thus always be a part of our conservation thinking, because the conservation choices we make inevitably shape the kind of people we are, whether we're attending to the ethical contours of those decisions or not.

Examining different candidates for what we should aim to conserve, whether its objects, memories, history, feelings, or authenticity, has revealed that there are always implications for individual and group identity lurking under the surface. Bringing the relationship between conservation and identity more firmly into view promises to influence how we think about what is worth preserving. But *what* we preserve doesn't shape our sense of self on its own. It does so in conjunction with how we answer the other conservation questions. In particular, there is a tight connection between our understanding of *what* to save and *why* it is worth saving.

3 FROM WILDERNESS TO POTTERY

WHY SHOULD WE SAVE THINGS?

Imagine you are standing in the attic of your beloved grandmother's house. Maybe you are stooped under the eaves; maybe you need to navigate along the joists where there is no flooring. Whatever differences our imagined attics might hold, it is hard to imagine an attic that isn't full of stuff. Imagine that your grandmother has recently passed and it has fallen to you to figure out what to do with the accumulations of a lifetime. Unless you have endless resources—space, time, money—you will have to make some choices about what to keep. How do you decide? Why save this rather than that?

The scenario you encounter in this imagined attic shares a surprising amount in common with preservation problems we face more broadly. All the world's an attic, and all of us, merely inheritors. If we had limitless time and resources, we could plausibly conserve anything anyone found some good in. If it floats your boat or tickles your fancy, provided there's no cost involved in hanging onto it, why *wouldn't* you save it? But faced with more restrictions on our capacity to conserve, we need to apply more stringent criteria—not just those things that someone happens to find pleasing or stimulating or useful, but just those that we deem worthy of such a response. Even then, the time, attention, and resources required for conserving all of the meaningful objects, activities, and places in the world easily outstrip our capacity for saving them. So, we find we need to raise the bar further. We find that in order to advocate for saving one treasured object as opposed to another, this tradition instead of that one, we will need a stronger justification.

The writer Wallace Stegner was a fierce proponent of wilderness preservation. In what came to be known as his "Wilderness Letter," a text invoked in the introduction of the Wilderness Act to Congress, he wrote: "We simply need that wild country available to us, even if we never do more than drive to its edge and look in."[1] Even if people can't get close enough to so much as peer into a protected wilderness area, he wrote that "they can simply contemplate the idea, take pleasure in *the fact that such a timeless and uncontrolled part of earth is still there*" (emphasis added). To be sure, Stegner catalogued a variety of reasons that humans can benefit from wilderness protection (among them the famous pronouncement that "it can be a means of reassuring ourselves of our sanity as creatures, a part of the geography of hope"), but in these lines on the mere contemplation of protected wilderness, we encounter the suggestion that wilderness is worth preserving because its continued existence is simply good.[2] Good in itself. The contemplation of preserved wilderness might also make us happy, but that's a bonus—it's the not the reason for saving it.

In his plea on behalf of wilderness protection, Stegner invokes an idea that is common in all kinds of conservation efforts. If conserving things involves according them a special kind of treatment, a level of care and concern that will be denied to other things, then perhaps those things worth conserving have a special kind of value that justifies this heightened regard. Maybe we should focus on preserving certain things because they are *intrinsically valuable*, as Stegner's Wilderness Letter implies. Or maybe we should save things because they are *irreplaceable*, and so failure would result in the loss of something we can never regain. We might also frame preserving things as a way of showing them an appropriate level of *respect*. All of these rationales, each advocated for by experts, has something going for it, but they also each face limitations as a complete justification for conservation. They are like independent pieces of a machine that hasn't been assembled yet—they each have an important function, but they aren't working in concert. We can address their shortcomings by building the machine, situating appeals to intrinsic value, irreplaceability, and respect on the organizing framework offered by the preservation of identity, like the parts of a car on a chassis. It turns out that maintaining and shaping identity, both individual and collective, is not only what conservation work achieves, as we saw in the previous chapter. It also offers a key part of the answer to why we should endeavor to save the things we value.

Intrinsic value travels under many guises—inherent, innate, independent, essential—but each label refers to a value that shares a common core. Intrinsic value is the significance that something has in itself, apart from whatever goods might be derived from using it. Looking at Monet's *Water Lilies* might cheer you up when you're in a foul mood, but that would be an instrumental value of the painting, the value it has as the means to achieving a particular aim (lifting your spirits). And while *Water Lilies* might be an excellent tool for cheering you up, that doesn't seem to capture what's distinctive or special about it. The intrinsic value of the painting, in contrast, is the value it has in its own right, apart from any particular use we can make of it.

Because intrinsic value isn't reducible to any instrumental contributions that something makes to the world, it can be hard to get a grip on, and philosophers have at times posited some fairly bizarre ways of thinking about it. For example, imagine a lonely planet, floating in space, devoid of life or even topography. Just a featureless sphere, and somewhere on the surface of that sphere—Monet's *Water Lilies*! Viewed in isolation from all other things, would it be good that *Water Lilies* exists? If so, then it has intrinsic value.[3]

If this "isolation test" strikes you as strange and uninformative, you're not alone. It can be difficult to get a grip on what it even means to imagine a product of human skill and creativity, such as a painting, in isolation from all other things. The problem with the isolation test is that it zeroes in on asking whether something has value in virtue of its intrinsic *properties*, the features that it has essentially, independently of its relation to other things. Alone in the world, we ensure that there is nothing in virtue of which the painting might be valuable besides its own features—its hues, its brushstrokes, its arrangement of forms, and so forth. But adopting this understanding of intrinsic value leads us to lose sight of the beings who might care about such things in the first place: namely, us! Especially when it comes to questions about why to conserve something, the matter of whether an object has value in virtue of its intrinsic properties (in isolation from other things) seems far less pertinent than the matter of *how* we ought to value it. Viewed from a practical conservation perspective, intrinsic value isn't an abstruse metaphysical description of the relationship between an object's value and its properties; rather, intrinsic value concerns the kinds of attitudes that we take toward an object.[4] To value something intrinsically is to view it as good because of what it is, as opposed to good because of what it can do for you. The philosopher Katie McShane describes how a range of fundamental human attitudes involve this kind of intrinsic valuing: "To respect something is in part to treat it as having a kind of importance in its own right; to be in awe of something is in part to treat it as having a kind of greatness in its own right, to revere something is in part to treat it as having a kind of dignity or nobility in its own right. One reason that we might find the concept of intrinsic value useful then is that we seem to do a lot of intrinsic valuing."[5]

This final point is worth stressing. Intrinsic value is a useful concept because it captures an essential aspect of human evaluative behavior. If we end up with a view of intrinsic value that makes it seem alien and strange (the province of Monets on empty planets), divorced from the everyday actions and attitudes of our lives, then we've likely missed the mark in our understanding of the concept.[6] To be intrinsically valuable in a sense with any relevance for conservation (or for our lives in general, for that matter) is a way for something to be valuable *to us*, not in isolation from us. But for something to be intrinsically valuable is also for it to call out for actions and attitudes that recognize its importance in its own right, apart from what it can *do for us*.[7]

While intrinsic valuing plays an important part in our lives, capturing a posture that we take toward a range of things we care about, the role it ought to play in conservation isn't straightforward. Awe, love, and respect may be attitudes characterized by intrinsic valuing, but that doesn't necessarily mean that intrinsic value is more important than other values, nor that the things we value intrinsically ought to be prioritized in conservation efforts.[8] Medical techniques, for instance, are paradigms of instrumental value—we care about them because of the effects that they produce, such as curing illness. Whether or not we also value these techniques intrinsically isn't really germane to whether we should preserve them: the importance of ensuring that these techniques are not lost or forgotten is that they save lives. *Water Lilies* might have intrinsic value, but that doesn't mean it's *more* worth preserving than the knowledge of life-saving medical techniques.

I suspect that the impulse to appeal to intrinsic value when advocating for the conservation of culture, art, and history stems from the perceived *lack* of instrumental value some of these things can have; or, in a more sanguine register, the *insufficiency* of instrumental value to explain what is special about them. *Water Lilies* doesn't necessarily *do* anything—it certainly doesn't save lives (though you can understand why preservationists might be motivated to wax poetical about an artwork's ability to do so). And even if you think great artworks can provide a particularly intense or rewarding experience, so long as we understand their value just in terms of producing a particular response, we leave open the possibility that anything else that produces that response would be just as good (maybe a virtual reality experience of *Water Lilies*, or a drug that induces the same feeling). So, whatever instrumental values we attribute to *Water Lilies* won't capture what's *special* about it, what makes us feel that it's worth preserving *this painting* as opposed to some other thing that can deliver the same goods. This point will be all the more salient for cases of things we value intrinsically that do not offer such experiences in the first place (an aesthetically unremarkable heirloom, for instance). The point, then, isn't that the purported intrinsic value of *Water Lilies* makes it more important or worth saving than the instrumental value of crucial medical techniques—it's that in the absence of an appeal to intrinsic values, it becomes harder to explain why saving *Water Lilies* is such a priority. Beauty and knowledge and artistry are all paradigms of things we value intrinsically; not because they aren't also instrumentally valuable, but because their usefulness doesn't exhaust their value, doesn't fully explain the particular ways that we care about them. But because appeal to intrinsic value alone can't determine what we should save, we need to place intrinsic value in a context that makes clear when and why it matters. As we will see, identity provides one such context.

—

This discussion of both *Water Lilies* and medical techniques suggests that an important reason to preserve certain things, regardless of whether they are intrinsically valuable, is that they matter in a way that other things don't. Put differently, a good reason to preserve things may be that they are *irreplaceable*. You might let a sponge grow grimy in your sink for much longer than you should, perhaps out of laziness or frugality, but ultimately you're going to toss it and replace it without a second thought. While we have good reason to avoid being wasteful, anything that can be replaced at little cost or inconvenience is unlikely to inspire us to mount a conservation campaign. "Save the used toothbrushes of the world!" said no one, ever. There are real worries that consumerism and planned obsolescence have fostered a culture in which we regard *too* many things as easily replaceable, but even an appropriate recalibration away from fast fashion and disposables still leaves us with a world in which many things admit of easy replacement.[9]

The problem with appealing to irreplaceability as a justification for conservation, though, lies in determining what things should actually qualify. Irreplaceability does offer an important reason in favor of preservation—if we fail to save something that is truly irreplaceable, then that loss is irrevocable. We lose something that can never be regained. But if we're too permissive about attributions of irreplaceability, we quickly encounter a *proliferation problem*: a scenario at the opposite end of the spectrum from profligate consumerism, where we regard too many things as irreplaceable, such that the category no long offers any useful guidance about what to save and what to let go.[10]

The proliferation problem rears its head quickly in the many contexts where we might appeal to history in an attempt to secure an object's irreplaceability.[11] The reason we should save this artwork, this artifact, this species, is because it is the result and embodiment of a unique history. We might make an exact replica of *Water Lilies*, one that is indistinguishable from the original, but it won't be the same because it will have a different origin, a different story, a different history. As we saw in the previous chapter, that doesn't mean replicas can't be valuable—but if we assume it's the historical features of the object that we're interested in, the replica simply can't get us those. So, because no contemporary replacement can capture these features, we have a compelling reason to preserve the original.

So far so good, but this kind of appeal to history is too general. After all, *everything* has a unique history, whether it's *Water Lilies* or the dirty sponge languishing by your sink.[12] Appeal to a unique history alone would have the startling implication that basically everything is irreplaceable, and we live in a world of constant irrevocable loss. But, of course, we don't actually think that the sponge is worth preserving, even though no replacement sponge would have its same unique history, and we don't in practice view the loss of every little thing as a tragedy. Someone who did so would appear to be making an evaluative mistake, finding too much value in the world, elevating everything to a treasured level that ultimately fails to distinguish between a Monet and a toothbrush. This indicates that we need something more than appeal to a unique history if we are to capture a sense in which irreplaceability really matters, a sense that could offer some kind of guidance when it comes to considering what is worth preserving and why. After all, the problem of preservation is thrust upon us precisely because we can't save everything. An approach that implies that everything is worth saving does us no good, as we can't possibly live up to it.

It turns out we don't need a story about what makes something unique; such tales come too cheaply. What we need is an account of what makes something uniquely *valuable*.[13] While every object has a unique history, not every object is historically *significant*.[14] This is particularly clear when it comes to items we cherish because of our personal histories, such as mementos, keepsakes, and family heirlooms. The connections between the historical features of these objects and the people we love makes those histories significant for us. I might have a passing interest in the production history of my grandfather's watch, but that has nothing to do with why it matters to me—I care about it because it was my grandfather's!

Of course, sometimes our obsession with personal artifacts can be "all too human" (as Nietzsche might put it), leading us to get attached to a quantity of keepsakes that outstrips the rationale for hanging onto them. Even when we do have good reason to view something as historically significant, it can still be the case that a bunch of things are historically significant in roughly the same way—historical significance doesn't guarantee irreplaceability. Put differently, when a class of objects is historically significant, that does not imply that every member of that class is uniquely valuable. This can be true whether the class of objects in question has personal or world-historical significance. Let's suppose that Ancient Roman pottery has some historical significance. We of course can't manufacture more Ancient Roman pottery—the overall supply is limited. In this sense, Ancient Roman pottery as a class of objects is irreplaceable. But the fact that Ancient Roman pottery as a class is irreplaceable doesn't mean that every individual piece of Ancient Roman pottery belongs in a museum. This doesn't imply that we should treat a piece of Ancient Roman pottery like a dirty sponge, but we also should be wary of views that imply (even if unintentionally) that every shard is an irreplaceable treasure that demands preservation. It turns out there's a ton of Roman pottery lying about (the Roman Empire was massive, folks—they needed a lot of plates and bowls), so treating each piece like a rare jewel quickly approaches absurdity.[15]

Appeals to irreplaceability invoke an absolute that can cast conservation in extreme, all-or-nothing terms, but many conservation questions are far more nuanced than such a concept would suggest.[16] We might reach for appeals to irreplaceability in order to play a trump card in a conflict over conservation, but as we have seen, justifying the claim that an object is irreplaceable will be far from easy. This is not to dismiss the relevance of irreplaceability to conservation, but to emphasize that we need to be discerning about it. As with intrinsic value, we'll see that by situating claims about irreplaceability in particular frameworks of meaning, such as those connected to identity, we can help to explain when and why they matter.

But before we get there, we need to talk about respect. We already considered the idea that valuing something intrinsically, for its own sake, is paradigmatic of certain central human attitudes, such as awe, love, and respect. It's thus not incidental that we often appeal to intrinsic value when we're advocating for conservation, especially when it comes to things we care deeply about. To identify something as intrinsically valuable is to say that it is *worthy* of the attitudes that are characterized by intrinsic valuing, such as respect. So, to identify a conservation candidate as intrinsically valuable is to offer a reason in favor of saving it—we ought to respect it, and conserving it as a way of doing just that.

The reason that conservation is supposed to express respect is often left implicit. We might try to forge the link between these two ideas by first considering the connection between their apparent opposites: neglect and disrespect. To lack respect for something typically involves spurning it, disregarding it, treating it as below one's care and concern. Practically speaking, this will often involve neglecting, if not outright disparaging, the things we lack respect for (the difference between neglect and disparagement might track the distinction between merely lacking respect for something and actively *disrespecting* it). But the fact that disrespect will typically involve neglect doesn't mean that respect will require conservation. Conservation may be a way of expressing respect for an object, but not necessarily and not always. For one thing, there are contexts in which forms of ritualized destruction are paradigmatic of what respectful behavior calls for. According to the U.S. Flag Code: "The flag, when it is in such condition that it is no longer a fitting emblem for display, should be destroyed in a dignified way, preferably by burning."[17] Tibetan Buddhist sand mandalas are swept away after they have been created, to express the Buddhist recognition of impermanence.[18] Wooden sculptures of the twin Zuni deities *Ahayu:da* "are to remain at their shrines exposed to natural elements until they disintegrate and return to the earth."[19] In all three of these examples from diverse cultural contexts, different forms of destruction are what respectful action calls for.[20] To insist on conserving a tattered flag, a sand mandala, or a sculptural representation of *Ahayu:da* would generally be to treat these objects inappropriately rather than to show them respect. The same reasoning can also surface in thinking about the preservation of social groups and practices, not just objects. When the rock group The White Stripes stopped playing together in 2011, they explained the decision as follows: "It is for a myriad of reasons, but mostly to preserve what is beautiful and special about the band and have it stay that way."[21] In this case, *breaking up* the band is seen as the route to saving what matters about it.

That being said, wasteful destruction and willful neglect do express disrespect for valuable things, not to mention something potentially vicious about those who disparage things in these ways.[22] In May of 2022, the TikTok account of a company called Engineered Labs posted a video of a person who appears to casually destroy a 3,000-year-old piece of Indus Valley terracotta by intentionally dropping it on the ground, allowing it to shatter.[23] In the video, the terracotta is clearly labeled and stored, and the person handling it is wearing latex gloves, cultivating the impression that they are expressing care and concern for this ancient artifact. These details only add to the shocking display of wanton destruction before us. Watching the video, we feel like we are spying on a museum conservationist at work, a trained professional handling an important artifact in their charge. And then they just let go.

Disrespect was a common theme in the outcry that followed Engineered Labs posting the video.[24] The company quickly apologized for the stunt and explained that the terracotta was a modern replica, not a piece of ancient pottery after all. But multiple commentators noted that the fact that the terracotta was a replica does not undermine the disrespect that the act expressed. They pointed out the links between the video and the extraction of Indigenous material heritage to Western museums through colonialism. According to one archaeologist, the video communicates to viewers that such disrespectful actions are acceptable.[25] On one reading of these criticisms, the concern is that the video will reinforce unjust relations between Western collectors and post-colonial communities, potentially leading to further harm by reinforcing norms of entitlement surrounding these objects—appearing to destroy ancient pottery on a platform that reaches millions of people just because you can is a power move, and the power concerned is power *over* those who have been deprived of possession and control of their material heritage.[26] But even apart from whether the video influences anyone else's behavior or leads to further destructive actions, we can also read the video itself as expressing disrespectful attitudes, regardless of whether the pottery is authentic or a replica. It embodies a cavalier posture, not just toward the object, but toward what the object *means*, and toward those for whom it is precious.[27]

It can be illuminating to compare this case of pottery destruction with another. In his 1995 work *Dropping a Han Dynasty Urn*, the artist Ai Weiwei presents a photographic triptych of himself intentionally dropping and destroying a 2,000-year-old ceramic urn. Unlike in the Engineered Labs case, this was a genuine piece of ancient pottery. But also unlike in the Engineered Labs case, Ai was incorporating his destruction of the Han urn into the production of a meaningful new artwork. Perhaps part of Ai's aim was to express a kind of animosity toward the relic, even disrespect. But it wasn't *merely* disrespectful destruction—it was destruction that sought to create another good (the artwork), that was dependent on the destruction for its meaning. Artworks—including pieces like Ai Weiwei's—are sometimes *transgressive*, requiring the violation of ethical norms in order to deliver an aesthetic punch.[28] Viewing Ai's work as such a deliberately transgressive artwork helps us see his act of destruction in a different context than that of the TikTok publicity stunt.

The point is not to insist that Ai's destruction of the Han urn was justified. Perhaps it was wrong for him to smash it, even in the service of his art—transgressive art is complex, and worth interrogating. But rather than presenting an unusual case of a boundary-pushing artist, Ai's destruction of the urn in fact illustrates a paradigmatic situation. We constantly encounter scenarios where we need to make trade-offs—destroying things or allowing them to decay in order to preserve or create other goods.[29] But there are ways of expressing respect for valuable things even when it's been determined that they cannot be saved. As the philosopher Joseph Raz has noted, one essential dimension of respect is simply according valuable things attention in one's thoughts.[30] The idea is that there is something distasteful about thinking ill of something worthy of respect, even if no one would ever know, just as there's something distasteful about wishing harm on your best friend, even if they'll never know you entertained the thought. The point isn't to be overbearingly puritanical about our psychology; if having vicious thoughts cross your mind is wrong, then cart me off to morality jail! But it does matter morally how we treat these thoughts on reflection. Do we endorse them or repudiate them? How we self-consciously regard something is an important dimension of whether we really respect it.

But moreover, internal attitudes of respect can find expression in the ways that we approach letting go of things that we are about to lose. Depending on the objects in question, this might involve opportunities for public displays of grief, or repurposing of materials from a destroyed object or site. While respect is an important value that structures appropriate responses to significant objects, practices, and places, it doesn't require that the things we value be preserved. Respect can be part of the reason in favor of conservation, but it can't be the whole story.

As we saw earlier, and as you no doubt knew before you picked up this book, we can't make new Ancient Roman pottery. This fact has led some scholars and institutions to begin referring to our material heritage as a "non-renewable resource."[31] While this is a well-intentioned analogy, its implications in the conservation context are unclear. In the case of other non-renewable resources, such as fossil fuels, we are in the process of using them up by intentionally destroying them at an alarming rate. To say that fossil fuels are non-renewable highlights that we are on pace to run out, and that we should prefer the use of renewable resources that are not subject to the same limitations (not to mention harms). But we're not at imminent risk of running out of Ancient Roman pottery, in part because there's quite a bit of it, but more germanely, because we're not in the business of intentionally destroying it. And even for heritage goods that are rarer, we're not looking to replace them with alternative sources of value. Most importantly, while certain historically significant objects might be non-renewable in the sense that we can't engineer replacements that would have the same history, this doesn't necessarily mean that their *value* is non-renewable. As we have seen, we tend to value many historically significant things intrinsically—they are objects of awe and respect, and we view them as warranting these attitudes. But our attitudes are not like perpetual motion machines; on the contrary, like real engines, they require maintenance and attention in order to be kept in working order. If we think it's important that some historically significant objects play a role in our lives, then we may well need to renew our relationships with them in ways that continue to cultivate the attitudes we think they merit, such as awe and respect. This idea is explicit in UNESCO's *Convention for the Safeguarding of Intangible Heritage*: "intangible cultural heritage, transmitted from generation to generation, is constantly recreated by communities and groups in response to their environment, their interaction with nature and their history, and provides them with a sense of identity and continuity."[32] It would be a mistake to assume this process will simply happen on its own. As philosopher Anthony Cross puts the point in discussing the preservation of shape note singing: "This preservation and cultivation constitutes a kind of *temporal* achievement: the successful collective endeavor of safeguarding the old songs from the vicissitude of culture while also keeping the value of singing them fresh and relevant for new generations of singers."[33] Relationships require work.[34]

A nearby error involves thinking that the kind of maintenance required to protect the role of these objects in our lives is material conservation only. Recent work in heritage studies has been at pains to show that even objects traditionally understood in material terms are immersed in a network of intangible practices, meanings, and relationships.[35] As we saw in the previous chapter, what we're aiming to save when we engage in conservation is not just the object, but our sense of self and the role that the object plays in contributing to our identity.[36] This becomes even clearer when conservation candidates are not physical objects at all, but traditions, customs, or languages, whose future existence is tightly bound up with their continued role in social life. But even for these "intangible" goods, it can be easy for our approach to their conservation to become assimilated to the case of physical objects. In Chapter 4, we'll look more closely at practices that can facilitate this broader form of conservation, and see that the lessons for approaching conservation across cases of the material and the intangible should generally be flowing in the other direction. For now, though, we'll focus on why this conservation matters.

—

The word "identity" has become a lightning rod in public discourse when it should really be a rain barrel. Rather than sparking contention through its mere mention, we should recognize its role as an essential bin for collecting and organizing the inundation of components that contribute to and constitute a life. Everyone has an identity that provides some shape and meaning to their lives. Identity, in this sense, is just a way of describing yourself that matters to you.[37] Whether you're a conservative or a liberal, a sports fan or a bookworm, a dog person or a cat person, you have an identity in this sense, and these examples obviously don't even scratch the surface of the available options. You might identify as a pescatarian, cinephile, mystery novel fan who's into competitive kite-flying—if that description captures what you value about being you, then that's your identity in the sense that concerns me here. So, as should be clear, this is not just the "identity" of "identity politics" (though it includes that notion, too). It's a much broader sense of the concept.

We can use this understanding of identity to help explain the sense that there are certain things that we feel we *ought* to do (leaving aside for the moment those things that we *morally* ought to do).[38] If you care about your identity, and certain actions will threaten that identity, then you will feel the normative pull that you ought to act so as to maintain the identity that you care about.

If you identify as a runner, for example, you'll be predisposed to do the things that runners do—make sure you have appropriate footwear (Or train for barefoot running? Is that still a thing?), adopt a stretching routine, make time in your schedule to go for runs, and so forth. If you find that you can't manage any time to go on runs, you may feel your identity as a runner start to fray: how can you be a runner if you don't even go running? Failing to do the things that runners do threatens you with the loss of that identity. If maintaining your identity as a runner is important to you, then you'll feel that you *ought* to take steps to prevent that loss, such as redoubling your efforts to find time to go on runs, maybe joining a running club to keep you accountable. This is not to claim that reflecting on our identities is a constant or self-conscious aspect of our decision-making—it would be psychologically unrealistic to claim that every time a runner decides they ought to go for a run they're motivated by an occurrent desire to maintain their identity as a runner. Sometimes it's just a nice day! Moreover, it would be a mistake to assume that the role identity plays in our understanding of what we ought to do renders it self-serving, even if it is in part self-focused. This would be an overly restrictive, navel-gazing conception of identity that doesn't track the wide variety of identities that are bound up with the welfare of other things, places, and people. Part of my identity is as a parent, for example, and my identity as a parent is entirely wrapped up in the welfare of my own child. I care about being a parent, but no remotely plausible conception of why I care about it could be disassociated in theory or practice from the other-regarding concern for my daughter. My identity as a parent offers a description that explains some of the things that matter to me, but there are a host of more specific concerns that make up that identity without making reference to it—much of what I do as a parent isn't *about* being a parent, though my identity as a parent *describes* many things that I do in that role. When my daughter is in pain, I don't pause and consider whether offering help and compassion is what a parent should do—I simply go. And that, in part, is just what it is to be parent.[39]

The idea, then, is that identity offers a framework for organizing and understanding the relationship between preservation and the valuable objects, practices, and places (not to mention people!) in our lives, because these elements all play a role in *constituting* who we are. Aspects of this way of thinking about identity as inherently relational can be found in intellectual contexts from Confucianism to the Nguni concept of ubuntu to contemporary psychology.[40] We do not become who we are in a vacuum, but in constant interactions with the world around us. Identity won't always be an operative part of our conservation reasoning, but it exerts influence on it nevertheless. And as we'll see, there are many conservation contexts in which it becomes an explicit part of the conversation.

It's important to see that the normativity that stems from identity, the sense that there's something that you ought to do in order to preserve your identity, is dependent on your caring about maintaining that identity. That identity has to be something that you value, such that the threat of its erosion motivates you to take steps to salvage it. I personally despise running, and the mere idea of identifying as a runner chafes against every fiber of my being. For me to fail to take the steps that constitute an identity as a runner is meaningless—I don't want that identity in the first place. But imagine a scenario in the middle. Say that you have identified as a runner for a long time (you were on the high school cross-country team, maybe you even ran a marathon once), but you're finding that you have less time for running in your life, and frankly, less enthusiasm. Perhaps you've started cycling instead, and you realize one day that you haven't been for a run in months. And you decide that you're OK with that. Being a runner just isn't that important to you anymore. In this case, there's no strong sense in which you ought to take up running again—the reasons you have to do what runners do depend on your wanting to maintain that identity, and that's up to you.[41]

We've been imagining scenarios where you could make time for running, but haven't. We've seen that if you care about maintaining your identity as a runner, you'll take steps to go running, but you also might decide that that identity isn't so important to you anymore. But what if you care about being a runner and the constraints on your ability to go running aren't stemming from your own decisions (you've been dedicating so much time to reading and gardening), but are being imposed on you by others? What if street harassment in your neighborhood has become so oppressive that going running outside has become untenable? In this case, others are threatening an aspect of your identity that matters to you, and that changes the equation. In a case such as this, where the threat to your identity is also *unjust*, the character of the threat to your identity as a runner is different. It shifts from being a matter of personal pragmatic decision-making to a situation that is morally inflected. It should be up to *you* whether to be a runner or not, free of unjust constraints on that decision. And this is true even if you don't already identify as a runner—the situation in your neighborhood forecloses on what should be a viable option for how to live your life.

This case captures how the preservation of a practice, even a practice such as neighborhood jogging, can play an important role in the ability of people to maintain or adopt a particular identity. While the scenario I've described is specific, the issues that it illustrates are generalizable and powerful. Any identity will involve constituent elements, whether they are practices, objects, or places. Threats to those elements of our identity, or our ability to engage with them in appropriate ways, can in turn threaten our sense of self. And this is ultimately *why* questions about conservation often matter to us so much. What's at stake is whether we are able to maintain some aspect of who we are.

The steps it takes to count as a runner aren't hard and fast (no puns intended), and there's a lot of room between the casual fun runner and the hardcore marathoner. But at the limit, at least, it's pretty clear—you need to do some running. For many other identities, however, what constitutes having the identity in question, and what people will feel as a threat to that identity, can be subject to considerable contestation and negotiation. Moreover, we don't just have a single identity. Your identity as a runner can sit comfortably with other identities you may have: parent, Celtics fan, cheese snob, Korean-American, libertarian, and so forth. Some of these identities might be more important to you than others; how you perceive the essential elements of some identities might be clearer for some cases than others. But whatever controversies surround our understanding of how to define a particular identity, it ultimately needs to have some constituent elements, some pieces out of which it is assembled. Caring about the preservation of that identity will in turn motivate the preservation of the components that make it what it is.

—

The relationship between identity and conservation only increases in complexity when we take into account that our identities are not only individual: they are also collective. We have identities as groups that range from interest groups to ethnic groups, religious groups to national groups. Some have claimed we even have an identity as a species, an identity that concerns how we understand what it means to be human. As group identities intersect and conflict, they complicate how we consider what is worth preserving, and they shape the contours of the components that can make up an identity.

Consider this comment on preservation of the Potawatomi language from Tesia Zientek, education director of the Citizen Potawatomi Nation in Oklahoma: "If you lose the language, we can still understand ourselves as Potawatomi. We, of course, still have our tribal governments, still have citizens. But we lose that way of looking at the world that is distinctly Potawatomi. And I think that would be an insurmountable loss."[42] Multiple concepts that we've explored in this chapter come together in this statement. The language is not only valued instrumentally, as a means of communication, but also intrinsically, for the distinctive perspective it offers that is itself a constituent of Potawatomi identity. The loss of the language would be *insurmountable*, another way of identifying a good that cannot be replaced. There are, as Zientek acknowledges, other components through which the Potawatomi could still recognize themselves as such, but to say that the loss of the Potawatomi language is insurmountable is, I take it, to register that while the Potawatomi would persist as a people, they would not be the same.[43] Situating the ideas of intrinsic value and irreplaceability in the framework of identities that we value and want to preserve (hooking them up to that chassis) helps make clear how these concepts matter to saving things in general. The reason that certain intrinsically valuable things are irreplaceable in a way that matters for conservation is that without them, we would lose an important aspect of who we are.

If one of the reasons for conserving things is to maintain our identity, it helps make sense of the felt normativity, the urge, to do so. But it doesn't always make preservation justifiable. Zientek's concern about the preservation of the Potawatomi language is justified if any case of preservation is. But appeals to the preservation of identity arise in all manner of cases, and they don't always hold the same weight. Until recently, I lived in Natick, Massachusetts, where there has been intense debate about whether to preserve a small dam on the Charles River, which runs through the southern part of the town. The current spillway was completed in 1934, with a complex history of prior damming activity, and some long-time members of the community have grown attached to the picturesque little waterfall that the dam has created.[44] Sure enough, in public debates about whether to preserve the dam, this attachment has shown up in identity-language. The sole Select Board member voting against removal described the dam as part of the "fabric of the community."[45] A citizen group dedicated to saving the dam describes it as "an integral part of the unique majesty of South Natick."[46] I'm pretty sure I saw a Facebook comment in a town group that pointedly asked *who we would even be* as a town without the dam. These local experiences echo broader academic work documenting the role that identity frequently plays in similar controversies elsewhere. The title of one such article features a telling quote: "You kill the dam, you are killing a part of me."[47]

Even when I was a fellow resident (though admittedly one that had not spent my life in Natick), I didn't find these claims to strong links between the dam and town identity compelling at all, but it's important not to dismiss them out of hand. Coming to see the way that other people view conservation issues as wrapped up with their identity is essential to understanding the contours of the debate, especially when you don't agree with them. What's at stake in deciding not to preserve a little dam in a small suburban town acquires a different character when you recognize that for some residents, the loss of that dam will involve a threat to their sense of self, at least in part. But the consequence of introducing an identity framework to conservation debates is not necessarily to offer extra impetus in favor of conservation. This is because maintaining an identity isn't always the trump card that advocates make it out to be. Sometimes we need to *reimagine* who we are. Recognizing the complex role that identity plays in motivating conservation efforts sets the stage for us to consider *from what* kinds of change we should aim to save things and *how* different approaches to conservation will hinder or help in the role of preserving identity.

It can sometimes feel like a stretch to claim that conservation is about the preservation of identity, but that rationale may be more or less proximate in any particular case, and it's not meant to exclude the many other factors that influence conservation decisions as well. To say that what we preserve when we preserve a painting is an aspect of our identity is not to imply that the painting itself, its aesthetic qualities, the creativity it exhibits, and so forth, are not also things worth saving—the aim isn't to render all conservation work instrumental labor in the service of maintaining identity. It's also not to say that when we approach the conservation of a painting we should *begin*, as a matter of practice, with questions about identity. But when you pull back the curtain, appeals to identity are typically lurking among our motivations, and as we have seen and will continue to observe, claims about identity are both implicit and explicit in preservationist rhetoric, even concerning our very identity as human. Consider this quote from Raphael Lemkin: "The contribution of any particular collectivity to world culture as a whole forms the wealth of all of humanity, even while exhibiting unique characteristics."[48] Human culture is the garden in which the many varied things worth preserving grow: a loss of any plant is a loss in itself, but it is also a loss to the garden. Lose too many plants, and it's no longer a garden at all. Who would we be without all of the spectacular and multifarious achievements of human culture? Something diminished. Something less than who we were.

It's worth pausing to consider how the link between human identity and conservation relates to contexts that can seem to resist this connection, such as conservation biology or environmental preservation. In what sense is the conservation of endangered species, for example, related to human identity? Aren't *we* generally the problem that species conservation is aiming to solve? Shouldn't it be about *them* (endangered species) and *not* us?

Some approaches to environmental conservation have a thoroughgoing anthropocentric cast to them.[49] They couch the rationale for the conservation of species and ecosystems in the ways such efforts ultimately serve human interests, even if the link isn't always immediately apparent. That's not the kind of approach that I'm recommending here. To say that identity sets a framework for approaching the conservation of the environment or endangered species is not to say that the primary or only reason to conserve these things is in order to maintain a certain identity or preserve our interests. As in the case of parenthood, I don't care for my child just *in order to* continue being a parent—caring for my child, for my child's sake, is what being a parent consists in. It's in fact *only* by attending to my child's interests for her own sake that I can maintain my identity as a parent (or at least a decent one). Likewise, it can be helpful to think about the ways that identity is linked with environmental preservation, but this doesn't reduce environmental preservation to an anthropocentric activity oriented around service to human interests. Various identities that help motivate conservation, from "environmentalist" to "land steward" to "urban gardener" are constituted in part by caring for the environment for its own sake.[50] But the environment is constantly changing—from what changes should we protect it? This is a question that surfaces in all conservation contexts, and how we answer it will have a profound effect on what we take the aims and methods of conservation to be.

4 FROM CLIMATE CHANGE TO COLONIALISM

FROM WHAT SHOULD WE SAVE THINGS?

All loss is change, but not all change is loss. What transforms a change into a loss? And how precisely should we understand the ways that loss matters? You can lose a library book and you can lose your life, but the significance of those losses are planets apart. The writer Kathryn Schulz discusses this unwieldy aspect of the term: "This is the essential, avaricious nature of loss: it encompasses, without distinction, the trivial and the consequential, the abstract and the concrete, the merely misplaced and the permanently gone."[1] A greedy term indeed. It gobbles up all the changes and spits them out as an undifferentiated lump that makes the word almost onomatopoeic—*loss*.

When we endeavor to distinguish among different kinds of loss, the *cause* of the loss plays an important role. Losing something to the passage of time is different from losing it to intentional destruction; a loss to ignoring is different from a loss to forgetting. Attention to the cause of a purported loss may in turn be the first step to recognizing that it is less a loss than a shift, a change that may leave us with something to mourn or remember, but perhaps also something to celebrate.

The term for the tendency of physical properties to break down over time due to entropy is "inherent vice," which carries the implication that such change is inherently *bad*. This is an instance of a more general phenomenon: when we identify a thing as *changing*, the very terms we use to classify that change conjure an understanding of the change's source and, ultimately, its meaning. As the philosopher Carolyn Korsmeyer frames the idea: "In short, cause is embedded in the very concepts of damage, deterioration, and degradation, which situate objects in relation to a set of external events." To employ such concepts "commands what we might term a particularly deep *narrative aura*, for it summons a story accounting for the damaged condition."[2] Damage might be caused by environmental conditions or a careless visitor at an art gallery, but to identify change *as* damage is by that stroke to invoke a general sense of the kind of cause, and how we are supposed to feel about it.

The idea that change is granted a certain character by the story it conjures points toward a role for us to play in shaping the nature of change. Part of this narrative aura comes not only from the cause of change, but from the terms on which a potential change is confronted, managed, accepted, rejected, and so forth. In other words, how we encounter change ends up shaping the change itself. This is why the terms on which we face change, and in particular the agency we exercise in doing so, are crucial to our understanding of that change's meaning, and so, in turn, the role it should play in our lives. Consider this idea in relation to the conservationist practice of collecting. Ahmir Questlove Thompson writes: "A collection starts as a protest against the passage of time and ends as a celebration of it."[3] One way to understand the process at play in this remark is that collecting involves changing the terms on which the passage of time is experienced. Trends and fashions change, and you collect to hold onto the marvels of a moment that is passing by. Clinging to the meaningful objects from a period in time becomes an act of resistance against change. But through that very act of protest, through changing the terms on which change is experienced, the experience of change itself is transformed—what began as protest is transmuted into celebration. After all, change, and the difference it produces, is part of what makes the things we care about special and distinct. When we can shape the terms on which we confront that change, the specter of loss can reveal itself as a source of appreciation.

Here's an organizing analogy from beyond the conservation context. Among the most significant potential changes that people can face in their lives is the prospect of becoming a parent. This can be true whatever one's attitude toward parenthood (yearning, dread, indifference, etc.) and whether or not one ever becomes a parent. Whatever the situation, the terms on which we confront the prospect of parenthood, and the ability to bring our own agency to bear on that possibility (or lack thereof), play a central role in shaping this encounter with potential change. Someone might desperately want to become a parent, but only on their own terms, when, where, and in the manner that they choose. Hence the availability of family planning resources will be key to enabling their control over when and how to become a parent. They might encounter any number of roadblocks toward becoming a parent, even with access to appropriate resources; but even when some things are *beyond* our control, being able to confront those eventualities on our own terms can shape what they mean to us and what we decide to do next. We can tell a parallel story about the person who *never* wants to become a parent, and the importance of being able to pursue that choice on their own terms. The point is that the nexus between the prospect of change and our ability to bring our agency to bear on that possibility will become part of the change's story, and hence shape its character. A change that would otherwise be welcome can be robbed of its desirability by compromised agency; a loss can more easily be overcome, reassessed, or perhaps simply survived, when we confront it on terms that are, to the extent possible, free.

We can make analogous points about the terms on which we confront change in the context of conservation. When we ask *from what* conservation should aim to save things, the answer is not a list of particular changes. Rather, it is from change on terms that would prevent us from exercising our agency in confronting that change, whatever it may be. In other words, it isn't necessarily change itself that conservation needs to manage, but the terms on which change is experienced. Understanding the shape of this claim will be our task in this chapter.

—

We have already encountered the notion of *aura*, and its place-based cousin *character*, in our discussion of what to conserve. Aura and character refer to the feel of things and places, and that visceral sense is often of a felt history, a story that wreathes something in an atmosphere of associations and connections. This is clear, as we saw, in the notion of *contagion*, and the way an object belonging to a celebrity or a relative can develop an aura through a shared history with them, soaking up significance through association. The narrative nature of aura also influences the way that aura has been wielded in defense of artistic originals in contrast with copies—it is a particular history of production, connected to the artist, that makes originals special.

A narrative is a way of organizing something's past, giving its path through time a direction and a shape that generates meaning. As Christina Riggs puts the point in the context of archaeological investigation of the past: "Archaeology looks for long-gone people in what they leave behind, but it is hard to get at why and how those things and places mattered long ago. A guitar is only a guitar, if you do not know who brought its strings to life."[4] It's situating an object in a context that generates significance. Given this function, it is no surprise that story or narrative has come to play an important role in conservation across diverse literatures and disciplines, especially in fields concerned with the natural and built environment.

On one model, we should think of the aims of environmental conservation in terms of what choices would best *continue* the narrative of a place.[5] We might think of this as analagous to discovering the unfinished manuscript for a story, and imagining we are tasked with writing the next chapter (if not completing it). What next steps in the story would fit with the found material? This approach paves the way for the idea that conserving narrative significance from the past to the future might allow for, or even require, physical alteration to a place.[6] In other words, we might need to facilitate *changes* to a place in order to conserve what matters about it. The philosopher Sven Arntzen contrasts "static preservation" with "dynamic preservation." Especially when conservation candidates are bound up with our sense of self, we might be motivated toward a form of static preservation that resists any and all changes, spurred by a fear that alteration will erode our sense of identity. Dynamic preservation, in contrast, is focused on the relationship between people and place, and how that relationship produces meaning. Allowing physical change might actually facilitate the maintenance of a relationship to place in a way that static preservation is ill-positioned to do.[7]

Practices and customs that include *destructive* elements nicely illustrate the idea that change and adaptation can facilitate conservation of significance or meaning (though they are often dissonant with commonplace approaches to conservation in the broadly Western tradition). Consider how some Shinto shrines in Japan are ritually destroyed and rebuilt every twenty years.[8] The preservation of such a practice from past to future is inconsistent with the conservation of the shrines themselves, and this approach to conceiving of conservation candidates was anathema to the formal commitments of international heritage organizations such as UNESCO prior to the Nara conference on authenticity, held in Japan in 1994. This is a case where the conflict between conservation perspectives primarily involved a lack of appropriate recognition from UNESCO, but other clashes more directly inhibit the maintenance of cultural traditions. As we saw in the previous chapter, wooden Zuni sculptures of the twin deities *Ahayu:da*, which are intended to succumb to the elements, cannot properly complete their life cycle when sequestered in museum collections.[9] Museums that retain these objects might think of themselves as saving the sculptures from decay, but to the Zuni, the museums are in fact impeding the objects' purpose. Not all change is loss.

On a narrative approach to conservation, the value of preserving certain objects or places stems from the way that they "embody" the histories and cultural identities of both individuals and communities.[10] In other words, it's not just that narrative helps us understand why these things matter—the things themselves are physical manifestations of those meaningful stories. But this approach is also complicated by the multiplicity of stories that can be embodied by any conservation candidate. That is one of the tricky things about narratives—they are inherently selective and perspectival. Think back to the controversy over conservation of the South Natick Dam. An important narrative thread in the dam dispute concerns the perspective of members of the Natick Nipmuc, a local Indigenous group, who explained that their relationship *to the river* was disrupted by the construction of the dam in the first place. The very name "Nipmuc" means "people of the fresh water." This is an example of what Kyle Whyte terms "vicious sedimentation," where the practices, customs, and relationships of one community are literally inscribed on top of another community's, as a dam is built on top of a river.[11]

If we want to conserve the narrative of the South Natick Dam, do we need to save the dam or remove it? It depends on which story we're attending to and whose it is.[12] This point also raises questions for approaches to conservation that emphasize the role of *continuity* in successful conservation.[13] The idea is that if conservation aims at maintaining something's integrity, then continuity, itself a narratively inflected concept, offers an important metric for assessing success. From one perspective, the removal of the dam would create a jarring discontinuity with the site's history. But viewed from the perspective of Indigenous relations with what is now called the Charles River, the construction of the South Natick Dam itself constituted an abrupt *discontinuity* for that place. To conceive of preserving the dam as offering a kind of narrative continuity disregards an important dimension of the site's story. In particular, we might think it picks up the narrative thread in the middle, ignoring what came before it. Narrative approaches to conservation are often intended to avoid overemphasizing particular moments in time, viewing conservation candidates in dynamic terms, but they are also subject to their own forms of time-bias. These biases can shape not only our understanding of the relevant narratives but of what the right conservation candidates should even be (e.g., the dam vs. the river).

Consider the way that philosopher Emily Brady explains the role of aesthetic *integrity* in environmental conservation. We have already considered Brady's helpful conception of aesthetic character in Chapter 2: a sense of a place's aesthetic distinctiveness that bubbles up from the component pieces that constitute it. She introduces the notion of integrity in order to offer guidance about what we should do when we face questions about conserving the character of a place. In keeping with criticisms of static preservation, she agrees that we should not conceive of a place's integrity in terms of an unchanging moment frozen in time, as if the landscape itself were like a photograph. Rather, she advocates for thinking about integrity in narrative terms, appealing to a trajectory that has a kind of "wholeness" or "continuity" to it.[14] While she contends that applying the notion of integrity will require evaluating conservation candidates on a case-by-case basis, she explains that this narrative approach will endeavor to be true to the site's history and avoid "sharp breaks in the narrative through change on a grand scale which creates incongruity and strangeness."[15]

If character picks out the distinctiveness of a sense of place, then it makes sense that conserving character would favor avoiding sharp breaks: the narrative approach allows for change, but a change that is characterized by a degree of continuity, lest the distinctiveness the character describes be lost. A forest will change over time, but you can't clear cut the trees and expect to retain what's special about that place. Writing in the context of building conservation, Thomas Yarrow notes that attributions of character by their very nature invite a "logic of continuity" that shapes understandings of both *what* can change and *how* it can change.[16]

We have already seen the role that character plays in directing the attention of many conservation fields, especially those focused on architecture, cultural landscapes, and heritage writ large. Policy documents, legislation, and the work and writing of heritage practitioners identify character as one of the primary targets of historic conservation. Character is a notoriously slippery concept, but one lesson that emerges in Yarrow's ethnographic work interviewing heritage practitioners in Scotland is that it is not a metaphysically intrinsic quality, but a situated one that emerges through the relationship between material and context.[17] For example, in making conservation decisions about a historical building, practitioners consider whether the building has undergone significant changes in the past as *part* of its character, and this conditions their thinking about *how* we should conserve it. Buildings with a history of addition and restoration are thought to have a character that incorporates such interventions, and thus opens the door to further additions in the future without abrogating the building's historic character; buildings without such a history have a character that resists intervention.

This practical approach to character underscores the ways that we are involved not just in conserving character, but *shaping* it through conservation decisions. Conservators do not stand outside of time, acting on an object with an independent history. Rather, their decisions become part of the historic character of the buildings, just as decisions about previous interventions did.[18] This doesn't mean that conservators should act as if they are unconstrained by the building's history, but it also highlights how decisions that diverge from a particular pattern of historical treatment might be better conceived as an addition or alteration to the character of a building, rather than destruction of it. This speaks to a conservation joke that Yarrow mentions: "there are two ways to destroy the historic environment—by not conserving it, or by conserving it."[19] On one reading, the joke works by appealing to a background assumption that all change is loss, whatever its source, whether time or the conservator's hand. But if character is a feature that we forge rather than find, there is less reason to embrace such a view.[20]

Brady casts the worry about loss resulting from discontinuity in terms of the production of strangeness—when a place's integrity has been compromised, it will feel strange to us. But while strangeness might be a common characteristic of discontinuity, it is also fleeting. The strange is unfamiliar to us, but we inevitably *familiarize* ourselves with what at first seemed strange the more time we spend with it.[21] The sudden removal of a large and dying tree from a familiar place can render that landscape shockingly alien, but that strangeness is softened by our repeated return to the site—the philosopher Arto Haapala writes that "familiarity may be nothing else but a visual habit."[22] This is not to dismiss the importance of continuity by inviting "you'll get used to it" as a legitimate response to radical disruption engendered by failures to conserve a place's character. But it does raise a question about whether it is strangeness and discontinuity themselves that are the problem, and not the terms on which strangeness is experienced. What are the conditions that brought about this strangeness? Who and what caused it? Discontinuity, like damage, invites a story about what brought it about.

—

These reflections remind us that narratives are not simply generated by the passing of time, falling from the printing press of history into our eagerly waiting hands. Not only are narratives selective and perspectival, but they "reflect, generate, and reinforce certain values and priorities."[23] When we choose to continue a settler narrative over an Indigenous one, for example, we make a statement about which history matters. We cannot appeal to narrative on its own to answer normative questions for us—this simply abdicates, not to mention obscures, the role that we play in making decisions about what and whether to conserve. Attention to narrative may facilitate maintaining certain threads of continuity. But which kinds of continuity? And why does continuity matter? Do we even want to maintain narrative continuity? We can't simply assume answers to these questions, or act as if appeal to continuing a story alone offers justification for our decisions.[24]

This point is highlighted by work on the relationship between narrative and sense of place. An operative assumption in much writing about the role of narrative in conservation, as we've seen, is that attention to narrative facilitates the continuation of the conservation candidate's character. However, as we've also seen, not all narratives deserve to be maintained—some narratives call out for disruption. Places are constructed in part through narratives, but this construction takes place in a context of unequal power. Consequently, places are often influenced predominantly by "master narratives" that are presented as "natural, unanimous, and eternal," and which dominate the story of the spaces that they govern (the use of the word "master" is not incidental).[25] These master narratives can be either upheld or challenged by "setting narratives," the symbols, structures, practices, and norms that are created through the built environment and its use. For example, consider how the persistence of Afrikaner street names in South Africa might provide a setting narrative that reinforces a master narrative about colonial power, and how changing those street names could help construct a new setting narrative that asserts independence from colonial influence.[26]

When I discuss these concepts with my students at Wellesley College, we conduct an exercise in which we develop lists of "master narratives" and "setting narratives" at the college (this is an activity you can do yourself in a range of contexts: your home, your town, your place of work). One of the first master narratives my students identify is always the idea that Wellesley is a women's college. Wellesley was founded as a women's college, and is institutionally identified as a women's college, but many students who have graduated from and currently attend Wellesley do not identify as women. Different setting narratives can serve to uphold or disrupt the master narrative that Wellesley is a women's college. For instance, ubiquitous institutional language that refers only to women and which employs only feminine pronouns upholds the narrative. But students engage in various forms of "critical place-making" that build setting narratives that disrupt this master narrative.[27] These include explicit and public rejection of gendered language (e.g., replacing "sister" with "sibling"), trans-positive guerilla art installations, or creating spaces on campus that welcome and affirm non-binary and transgender students.

The point is not that notions of continuity have no place in discussions about what it means to preserve the identity of Wellesley College. We might helpfully use notions of continuity to partially explain why we might conceive of Wellesley College as a "historical women's college" (acknowledging its history and its evolving relationship to gender diversity) rather than simply turning it into a typical co-educational institution, though such a decision could be equally well explained by appeal to contemporary considerations of justice and the value of a mission-driven institution committed to the education of those who are marginalized on the basis of their gender identity. But when it comes to interventions that are by their nature *disruptive*, some opponents of change will find it easy to fall back on appeals to "continuity," preferring a kind of conservatism to the "strangeness" that might attend radical change.

It should be obvious that a good conversation about how best to preserve a college's identity cannot be conducted independently of the individuals who make up the college community. This suggests an avenue for navigating the terrain of conservation and narrative that recognizes a role for continuity, but shifts from an exclusive focus on the conservation candidate to the relationship between that candidate and relevant communities whose identity is bound up with it. This is not to advocate for a simplistic policy of majority rule or to suggest that community preference is all that matters; but it is to insist that the community, in all its complexity, has an explicit role to play.

Recall Whyte's concept of *collective continuance*, the ability of a group to *self-determine* how to adapt to change while avoiding harm.[28] Collective continuance, as an approach to conservation, explicitly acknowledges that identities are plural and changing, and places value on peoples' ability to adapt to change on their own terms.[29] For example, many Inuit seal hunters now use gas-powered sleds. While this may be a change from pre-colonial hunting practices, it is one that many Inuit have accepted as an aspect of the contemporary practice.[30] To claim that the use of gas-powered sleds in the Inuit seal hunt is "inauthentic" (especially if this claim comes from cultural outsiders) would thus be to misunderstand the relation of the practitioners to the change, imposing a static understanding of authenticity that is itself characterized by abrogation of autonomy. It would be as absurd as claiming that it is inauthentic to shop at a grocery store rather than farming your own food. Conservation decisions that facilitate the collective continuance of group identities might sometimes seem to involve significant discontinuities if we focus only on the stories of conservation candidates themselves. But when we reframe our thinking about concepts such as *integrity* and *character* in relation to identity, rather than only the objects, practices, and places we seek to conserve, we also reevaluate the significance of continuity to conservation efforts.

Many identities, both individual and collective, are bound up with culture, and it is a truism about culture that it is in a constant state of flux. Culture is etymologically linked to *cultivation*, to the work required to allow something to grow and flourish. Culture is *alive*, and so any attempt to preserve culture (and the identities bound up with it) will need to be responsive to culture's mutable nature.[31] Pressed flowers can be lovely—that doesn't make them a substitute for living ones. But the fact that change is inherent to culture does not mean that cultures should simply welcome any and all changes.[32] As we saw in the example of the runner subject to street harassment, changes to practices can be unjust, something to be resisted rather than simply accepted "because cultures change." Put another way, we might think the kind of change that is consistent with cultural preservation needs to be characterized by a certain degree of *autonomous agency* by participants in that culture.[33] This is also linked with a persistent desire for *authenticity*: if change is managed with a high degree of agency, people will have more confidence that they are being true to themselves within the constraints that they are navigating. It's far from clear that being true to oneself in this sense has much to do with the conceptions of continuity that tend to arise in conservation thinking. Autonomous change might be characterized by a kind of continuity insofar as it is freely chosen, but one might autonomously choose to head off in a radically new direction characterized by stark discontinuity of a different kind.[34]

The way we exercise our agency in making conservation decisions thus shapes and is shaped by the things that we conserve—the way we treat a conservation candidate *now* ends up conditioning the ways it is appropriate to treat it in the future. This can have surprising consequences: once we embark on the conservation of an object, that very conservationist orientation can end up influencing our relationship to the object in question. To conserve something is to change it, not just by the physical interventions or manipulations that conservation might involve, but through the nature of our relationship to it. When we engage in conservation, we take on a role as caretaker that filters future interactions with the object of conservation. The nature of that role is complex, and can involve all manner of hues and distortions, bringing its object in and out of focus in different respects. To employ a different analogy, consider becoming the guardian of a child. What it means to be their guardian is far from fixed—it can shape your relationship in all manner of different ways. But when you become a child's guardian, you can no longer relate to them as if they were *just any child*. Your role conditions the kinds of relationships that are possible, attainable, and required. You have obligations to your child that you don't have to just any child; consequently, you can fail them in ways that you can't fail just any child. I hope that I can be my daughter's friend, but I know I can't *just* be her friend, because I am also, always, her father.

Taking on a custodial relationship to an object via conservation can have surprising consequences. In a fascinating ethnography of classic Mustang owners, the authors explore how the authenticity of the car is simultaneously a joy and a burden for the owners.[35] The cars need to be driven in order to qualify as automobiles and to play a valuable role in the owners' lives (and, in fact, to maintain certain parts), and driving them creates wear that in turn requires further maintenance. Maintaining the authenticity of the car is itself central to the owner maintaining their own identity as a classic Mustang owner. The authors' main goal in the study is to illustrate that the authenticity of a classic car is not just a symbolic value to be consumed—nothing more than cultural capital to be leveraged—but also places *constraints* on the owner's actions as a function of their attachment to the cars. Attachments are double-edged—they can both expand and limit our agency by shaping what we have reason to do.

The philosopher Monique Wonderly develops this aspect of attachment further. She describes how attachments generate a certain variety of *felt need*—the desire to engage with an object, to view it as important, to suffer when it is damaged.[36] Felt needs can both constrain and bolster our *agency*.[37] In the case of the classic Mustang owners, their attachment to their cars creates felt needs that place external constraints on their agency: maintaining the authenticity of the car ends up directing how they spend their time, money, and attention. But attachments are also a source of security that can help orient us toward the world and "reflect something deep about who [we] are as agents."[38] Caring for their cars grants the classic Mustang owners a sense of purpose—they endorse how their attachments organize their actions and attitudes. Wonderly explains that felt needs are also generated by what we *care* about, and these needs are typically bound up with caring about something for its own sake, such as the caring involved in loving something or someone. But the felt needs deriving from attachment, in contrast with caring, have a self-regarding dimension to them—what I am attached to reflects, at least in part, what *I* need to engage with, what *I* need in order to feel secure.[39]

The idea of felt needs deriving from both our cares and attachments offers rich resources that we can use to add further depth to our understanding of the relationship between identity and conservation. We have already seen how objects, practices, and places become bound up with our identity, such that threats to those things constitute threats to our sense of self. This can generate felt needs to save those things in order to maintain our identity. We can now characterize our relationship to these objects, practices, and places as taking (at least) two forms that will typically travel together: what we care about, and what we're attached to. Our caring captures our other-regarding concern, the way that we value things intrinsically, for their own sakes. Our attachments capture our self-regarding concern, the way that we desire to engage with things ourselves, and feel threatened by their loss. Those two dimensions of our relationships often work in tandem with each other, and help characterize how the things that matter to us are both important for their own sakes and important to who we are, thus generating the sense of intrinsic value and irreplaceability that we discussed in the previous chapter.[40] To the extent that we can exert control over our cares and attachments, the constraints that they create are simultaneous ones that we *choose*—we bind ourselves to objects, practices, and places, and these links are the materials that help to construct who we are.[41]

Of course, not all these connections to the world are purely chosen. Some are handed down, nurtured in us, or part of a group identity that we are raised to embrace. Remember that the link formed by attachment runs in both directions: we might sometimes feel that the things we are attached to belong to us, but we might equally well say that we belong to them. This aspect of attachment is made particularly salient in some Indigenous notions of connection to place. Krushil Watene writes that on the Māori worldview, people conceive of themselves as belonging to the land, an idea that can be found across multiple Indigenous traditions. She continues: "In such a view, it is difficult to accept that particular land, waterways and natural resources, so fundamental to identity, can be substituted."[42] When it comes to loss of place, then, there is the visceral sense that you are robbed of a place where you belong, and which is part of who you are.[43]

In recognizing the power of felt needs arising from our cares and attachments, it is also crucial that we remember that an explanation of the relationship between our sense of self and the motivation to conserve is not necessarily a justification of it. We need to first ask if the identity in question is worth preserving. It may well be that white supremacists feel the need to conserve public symbols of the Confederacy due to a felt erosion of their identity—the link between identity preservation and object conservation helps *explain* their motivation. But that on its own doesn't *justify* such conservation. For a justification, we need to look to the broader moral and political significance of the identity in question.

What we save in public (and with public funds) is especially important in this regard. It shapes the story we're telling about what society should care about. Conservation is not only guided by narrative considerations, but it also generates narratives, and the values reinforced by those narratives can become embedded in the landscape. A striking example comes from the case of Uluru-Kata Tjuta National Park in Australia, also known as Ayers Rock.

Settler Australians have visited for decades, often climbing the rock, an activity that the Aboriginal Anangu people view as inappropriate. Through a promising joint-management plan between the Anangu and the federal government, a substantial educational program consisting of signs, brochures, websites, and other guidance was developed to communicate that climbing the rock is inconsistent with *Tjurkurpa*, or Aboriginal law. And yet these efforts were partially counteracted by the way that park infrastructure can communicate a contradictory message: for instance, the main parking lot is at the foot of where the climbing route begins, depositing visitors where they will see "the evidence of ceaseless colonial habits sketched across the spine of Uluru in the shape of a singular scar."[44] To employ a concept introduced earlier in the chapter, there is a "setting narrative" built through these features. This case illustrates how narratives and their reflected values can literally be written into place, conditioning understandings of what actions and attitudes are appropriate.[45]

The philosopher Lea Ypi argues that the fact that a community is attached to some place or object doesn't ground any particular claim to it, but when a community is subject to structural injustice, their attachment can help determine *what* they are owed as a matter of reparative justice.[46] So, in the case of Uluru, the Anangu are attached to the rock as a sacred site and settler Australians are attached to it as a symbol of national pride and a recreational climbing destination. If we *only* attend to the fact that both groups have attachments to the site, it may not be clear how to resolve the conflict. But if we instead look to the structural injustice faced by Aboriginal Australians in contrast with settler Australians, we can see their various attachments not as explaining what makes their claims to the site matter morally or politically, but rather fixing the content of what is owed to those who have suffered structural injustice. It's the injustice faced by the Anangu that gives weight to their claims for remedial justice, and their attachment to Uluru identifies it as a site for that justice to occur.

—

Attention to the moral and political landscape of conservation can reveal contexts where the decision *not* to conserve can aid in the maintenance or crafting of identity as well. In an article aptly titled "Colonial Ruins Are a Fitting Epitaph for the British Empire," novelist Chibundu Onuzo describes how the decision to let colonial buildings go to ruin, from Myanmar, to Nigeria, to India, can be part of a post-colonial nation's approach to confronting the past.[47] To allow the material manifestations of colonial rule to fall into disrepair is a way of expressing an attitude toward this history of oppression and its ongoing legacy; it can be interpreted as a way of noting the ongoing imprint of colonialism on these contemporary nations by allowing the buildings to stand, but adopting a critical attitude toward their impact by letting them fall apart.[48] Or consider the activists in the United States who have pulled down or defaced statues of Confederate soldiers.[49] To reject public symbols that maintain an identity bound up with injustice is made meaningful in part by the refusal to comport with familiar conservationist approaches—the abrogation of preservationist norms surrounding statues makes the expressive power of the actions all the more salient.[50] For a similar point in a completely different context, the author Maurice Sendak once told a story about sending a drawing to a child, whose mother reported that the child loved the drawing so much that he ate it. Sendak recounts taking this act as a great compliment.[51] Sometimes unconventional ways of expressing appreciation are inconsistent with familiar modes of preservation, and gain their expressive power in part from transgressing them. It's because eating the drawing was an unconventional expression of love that its destruction is ultimately interpreted by Sendak as a high compliment. The valence of destruction can't always be read off the act alone.

Even when no actual damage is intended, the specter of destruction can be leveraged in support of a cause. In the fall of 2022, activists from the organization Just Stop Oil engaged in a series of actions targeting fine art in major museum collections.[52] They glued themselves to frames and splattered the glass enclosures of Vermeers and Van Goghs with mashed potatoes and canned tomato soup. As the protestors made clear, their goal was not to actually damage the artworks. On the contrary, it's the very fact that they value the artworks that inspired the activists to use them as a site of protest in order to draw attention to the costs of failing to take immediate and radical action in response to the climate crisis. Phoebe Plummer, one of the protestors who participated in the tomato soup action with Van Gogh's *Sunflowers* at the National Gallery in London put the point this way: "it's a beautiful work of art and I think a lot of people, when they saw us, had feelings of shock or horror or outrage because they saw something beautiful and valuable and they thought it was being damaged or destroyed. But, you know, where is that emotional response when it's our planet and our people that are being destroyed [?]"[53] To engage in actions that even challenge the *appearance* of conservation within a space that is explicitly dedicated to conservation work sends an especially impactful message: look at this whole infrastructure dedicated to the preservation of a precious and fragile thing! Where is that concern for the environment? Attention to how we regard the entire apparatus of conservation can thus be used as a tool for challenging received values and priorities.[54]

To be sure, the powerful mechanism through which decisions to conserve, or neglect, or destroy can serve to maintain or challenge values and identities cannot be morally evaluated in isolation. When the Taliban blew up the Bamiyan Buddhas, they, too, were using destruction to strike a blow against a certain conception of Afghan identity. That action in turn frames the question of how and whether the Buddhas should be restored.[55] Intentional suppression of culture through destruction can make the threat to identity especially salient, generating robust conservation efforts in response.

This link between conservation and identity in the context of armed conflict is effectively illustrated by the international conversation surrounding the destruction of cultural heritage in Syria during the 2010s. Both ISIS and the Assad regime engaged in destructive actions, with particular international media attention focused on ISIS's targeting of the ruins at Ancient Palmyra. As many commentators have noted, there are concerns that the focus on Ancient Palmyra displays both a Eurocentric bias in international attention (the ruins are of Roman origin) and an outsized preoccupation with material heritage relative to the massive loss of life during the Syrian Civil War.[56] Given that armed conflicts always involve loss of life, it is a common refrain that focusing on the destruction of buildings or monuments while people are dying is perverse. It is this criticism, in combination with the intentional targeting of built heritage, that thus motivates some world leaders to articulate a strong link between efforts to save material heritage and efforts to save lives. For example, former president of France, François Hollande, asked: "What is more important, saving lives or saving stones? In reality, these two are inseparable."[57] Former Director-General of UNESCO Irina Bokova stated: "Defending cultural heritage is more than a cultural issue—it is a security imperative, inseparable from that of defending human lives."[58]

These claims about the inseparability of saving heritage and saving lives have both pragmatic and theoretical dimensions. Pragmatically, they invoke claims about military strategy and the permissibility of devoting resources and risking lives with the goal of preventing the destruction of material heritage.[59] But quite apart from whether military intervention that aims to protect cultural heritage can be justified, these claims also invoke a tight link between the preservation of material heritage and the maintenance of identity. The idea that attack on manifestations of a culture is itself an attack on a people seems to be the animating idea behind such destruction in the first place, and thus triggers a strong counter-response.

Perhaps one of the clearest pieces of evidence for the important link between cultural preservation and identity is how frequently oppressive powers have labored to stamp out, suppress, and erase culture in order to destroy or subjugate a group of people.[60] In such contexts, the reasons for conservation that stem from identity will be especially powerful. If terms such as "deterioration" or "decay" conjure stories of natural causes for damage, "suppression" and "erasure" summon intentional ones. When change is imposed against our will, conservation is a way of exercising autonomy—inserting our agency into change that was impelled by forces beyond our control. But to act in the face of natural or unintended changes has a different meaning from acting in the face of intentional destruction. An object that I cherish might deteriorate over time, but its ultimate disappearance would not necessarily be an *affront* to me in the way it would be if you maliciously destroyed it. The meaning of intentional destruction, too, is shaped by context, including the motivation for the action and who performs it. The fact that tomorrow is the designated time for the ritualized destruction and rebuilding of a Shinto shrine would not license a passerby to frivolously light the structure on fire today, and we would have good reason to save the shrine from this fate, its imminent demolition notwithstanding. It is difficult to make sense of these important distinctions if we focus on the conservation candidate alone. Rather, it's the broader context of meaning within which the candidate is couched that imbues the source of change with its particular significance. What the source of the change means *to me* or *to us* as the people who care about the threatened thing will shape how it makes sense to respond.

For example, the way that "natural" changes to a place are driven by climate change can affect the significance of those changes for particular parties, and thus shape the meaning of conservation efforts. We know that the causal drivers of climate change have historically flowed from industrial activities in the West. Thus, threats to places or culture that stem from climate change have a fundamentally different meaning for communities that have already been subject to other forms of cultural eradication by the West, even if climate change is not intended as a tool of cultural control. It takes on a different meaning from threats wrought by nature alone.[61]

When a change or loss is the result of intentional destruction or suppression, it can thus lend moral force to the prospect of conservation. For example, commenting on a collaborative language revitalization program between Fort Lewis College and the Southern Ute Indian Tribe, Jenni Truillo explains: "Languages in the boarding schools were stolen. People were beaten, they had their skin scrubbed, they had their mouths washed out with soap, they had their hair cut. Because of the fact that language was stolen, it's our educational and even moral responsibility to revitalize these languages."[62]

Perhaps, then, the very fact that something was stolen or threatened gives us special reason to recover or save it. The philosopher Chike Jeffers has advocated for the importance of preserving cultural traditions connected to racial identities as a form of resistance against racial oppression.[63] In a context where the forces of colonialism and its attendant racism sought to erase, devalue, and degrade the cultures of colonized peoples, the preservation of those very cultural traditions takes on a particular salience. It need not embrace a static form of preservation, and Jeffers urges the importance of turning a critical eye on cultural traditions that may themselves involve injustices. Nor is the aim of cultural preservation in the face of colonialism to provoke fights over authenticity and police the boundaries of cultural essences.[64] But to celebrate culture in the face of its attempted degradation is to take a powerful stand against what is, at heart, an attack on one's own person. The act of working to preserve a cultural tradition can thus be seen not as an attempt to arrest any and all change, but a way of resisting certain sources of change; namely, those that stem from an effort to destroy culture in the name of domination. It isn't change per se that we should save things from, but change experienced on terms that would control our ability to confront that change.

The value of cultural preservation in the face of racism and colonialism can be found across the many contexts in which people are marginalized and oppressed simply because of who they are. As I write, conservative legislatures across many states in the United States are working to enact bans on public drag performances. This is part of a concerted effort to attack and erase trans people, and despite the fact that many drag performers do not identify as trans, the negative impact of the ban on the queer community writ large (trans or not) is no doubt part of the goal.[65] The same laws that would prohibit a man from performing in drag in public could be leveraged against a trans woman who is simply running errands because the state refuses to acknowledge her gender identity. That the attack on culture here is an attack on people is so thinly veiled it belongs in the story of the emperor's new clothes. If anything, then, these threats to culture generate extra reasons to preserve threatened practices as demonstrations of self-respect in the face of degradation.[66]

The subtle tension at the heart of cultural preservation as a form of resistance, though, is that it will end up changing the culture anyway. Dedicating oneself to the maintenance of cultural identity in the face of oppression promises by that very act to shift the meaning of that cultural identity, shaping it with new moral contours. These shifts can even generate new forms of identity, such as pan-ethnic racial identities oriented around resistance to racial oppression, identities which are themselves controversial and contested.[67] But when we embrace the idea that all preservation should be dynamic—not just the preservation of things in the most obvious state of flux, such as landscapes, but also paintings, and recipes, and buildings, and languages—then the worry about such changes diminishes. Like lineages that branch and bifurcate as organisms encounter new environmental conditions, everything worth preserving will inevitably change. It's the terms on which we encounter, embrace, reject, or manage that change that matter.

We have seen that when we consider the ways that conservation is connected to identity, worries about loss have particular salience. If to lose something is to lose who you are (in whole or in part), then we can better understand the deep commitment that people can develop toward conservation. Who will we be without our language, our homeland, our culture, and so forth?

The implicit answer to the rhetorical question "who will we be without X?" is *no one*. But while that may be a real fear, it's not the real threat. We fret about the loss of traditions, landmarks, and languages because we worry that without them we will be no one, when in truth, we will be *someone else*.

Perhaps that notion may be even scarier at first, but it also points toward the promise that loss can be recast as change and renewal. To be no one is to be a lack, a nothingness, a life without drive or meaning or purpose. If identity is a description of what you value about yourself, then to have no identity is to view yourself as worthless. But if change to the things we value results not in the unequivocal loss of identity, but rather, the creation of a new identity, then there is hope latent there: the promise of new meaning, new purpose, new sense of self. This doesn't mean that all loss can simply be recast as change, or that all loss is morally equivalent. But it does alter the equation. And it underscores the pivotal role played by *how* we pursue conservation, who is able to participate in conservation projects, and on what terms.

5 FROM LANGUAGE REVITALIZATION TO DIGITAL REPLICATION

HOW SHOULD WE SAVE THINGS?

How do you save a crow? Not this crow or that crow, but a particular kind of crow, a species. Were it not for the interventions of conservationists, the Hawaiian crow, or ʻalalā, would likely be extinct.[1] As it is, only a small number of individuals from the species still exist, and only in captivity. But have ʻalalā actually been conserved? Views differ. On the one hand, you might think that the captive ʻalalā are the same as their recent wild ancestors in the sense that matters: they are their direct genetic descendants. On the other hand, ʻalalā do a lot of environmentally situated learning from other members of their species, and conservationists have observed profound losses in the vocabulary and behavior of the captive crows. This is a primary contributor to the lack of success conservationists have confronted in attempting the reintroduction of ʻalalā to the wild. According to some commentators, the captive ʻalalā are simply not the same species anymore—despite their shared phylogeny, their lack of a shared "culture" with wild ʻalalā makes them something different, not authentic ʻalalā at all.

But this essentialist understanding of ʻalalā identity might also be thought to freeze them in time, turning them into museum pieces rather than dynamic, living, evolving organisms whose nature is constituted by more than their genetics and also isn't fixed—rather, what it is to be an ʻalalā is the result of the bird's relationships with a particular environment and its other inhabitants. Instead of thinking that the captive ʻalalā either are or are not the same as their wild forebears, we might say that what it means to be a real Hawaiian crow is itself dynamic, something that is being constantly renegotiated through the interactions between organism and environment. In this regard it seems that captive ʻalalā are like the Kansas City tacos from Chapter 2—they may not be the same as their forebears, but that doesn't make them something false. They are a version that is not unrelated to what came before, but shaped by different conditions: both limitations and possibilities. By navigating change, they have emerged as something distinct, but not as something worse, and not as something altogether different. They are part of a lineage that retains core features of earlier iterations, while also adding something new.[2]

In many contemporary conservation fields, from historic preservation of buildings, to landscape management, to fine art conservation, there is acknowledgment that conservation is fundamentally about *managing* change rather than arresting it, though the impression of the latter aim can be persistent.[3] Whereas the goal of simply trying to keep things the way they are can allow us to avoid some philosophical questions about conservation (at least up to a point), the recognition that conservation is about the management of change throws the door wide open to questions of value, meaning, and judgment. To manage is to direct, to control, to exercise authority over, but also to treat with care. Answers to what we should try to conserve and why will naturally lead to further questions about how we should negotiate the management of change. We have already seen that conservation questions are often helpfully understood in relation to issues of identity (sometimes front and center, sometimes in the background), and that the goal of maintaining aspects of our identity can give us good reason to conserve valuable objects, practices, and places that constitute important parts of who we are. We have also seen that the terms on which we confront change will play an important role in how we understand, evaluate, and respond to those changes. So, when it comes to the matter of managing change, an essential issue will concern *how* we conserve things in ways that allow them to play their distinctive roles in our lives, that allow for us to hang onto the parts of our identity that motivate us toward their conservation in the first place and confront changes to them on terms that we can accept.[4]

As we have seen, the role that conservation candidates play in making up aspects of our identities is often indirect. We don't merely value these things *as* parts of our identities (though that may also sometimes be true); rather, we value them in all manner of specific, distinctive ways, and it is precisely *because* of the important roles that they play in our lives that we come to see them as valuable parts of our identity in the first place. For example, many people value particular cuisines as components of their cultural identity, but there is a complex set of elements that allow for this connection. We care about how the food is prepared, whom we cook it with, the occasions it celebrates, the stories it tells, and, of course, how it tastes—we cannot preserve the role of cuisine in our identities without attending to these kinds of features.

So, while the role of conservation candidates as constituents of our identities sets a framework for understanding how to conserve them, it does not provide simple or straightforward answers. It does, however offer a sense of direction and a set of guard rails that can help guide us toward what successful conservation—successful management of change—will look like.

Given the role that identity plays in motivating conservation efforts, it should come as no surprise that approaches to *how* (and whether) to pursue conservation are themselves often explicitly couched in identity claims. In ethnographic work on building conservation in the United Kingdom, Thomas Yarrow interviewed a homeowner who said: "That's the thing with old buildings. I feel like I'm a custodian of it, rather than an owner who can just bend it to my will and to hell with the consequences."[5] To identify as a *custodian* rather than an owner is to adopt an identity that in turn shapes how you ought to approach your relationship with something, whether it's a building, an artwork, or a landscape. Failure to act in the ways that custodians do would threaten the self-ascription of that identity: you can't remain a custodian if you don't care for something in certain ways. A conservation officer is described by one interviewee as "an advocate" for a building, a further identity that can shape your approach to conservation, perhaps in subtly different ways from that of a custodian.[6] To be sure, what actions count as those of a custodian or advocate will remain underspecified, subject to context and contestation. But these examples illustrate another explicit way that identity sets a framework for thinking through how conservation should be approached.

In the case of material conservation candidates, it's natural to think that what they're made of will play a determinative, or at least influential, role in how they ought to be conserved. There is a general disposition among certain lines of thought to assimilate the object to its material. This is the kind of conceptual move we saw in controversies over protecting tangible heritage in armed conflicts, where dismissive remarks about "saving stones" reflect a conflation between a monument or sculpture and its material, rather than another status it might have (art, heritage, culture, and so on). To assimilate a sculpture to its material is to look at a complex, meaningful object and understand it primarily in terms of its substance. Might we turn our attention in a different direction?

Many scholars have tried to shape the way we think about change by shifting the timescales on which we look at an object, using change over time as a wedge for opening up our understanding of what an object even is. Tim Ingold likens buildings to processes rather than products.[7] In a similar vein, Fernando Dominguez Rubio urges us to see that a painting such as the *Mona Lisa* is not a static object, but a "slow event."[8] According to his view, a "thing" is an unfolding material process, whereas an "object" is a position in that process that affords certain understandings of meaning and value.[9] If we think of the *thing*, the *Mona Lisa*, in terms of the material process that began with its creation and ends with its eventual disintegration, then the *object*, the *Mona Lisa*, is the portion of that process where the painting is in a state that allows us to appreciate its artistic qualities. Conservation, then, aims at maintaining that sweet spot, where the position of the candidate at a certain moment in the process of its transformation affords particular meaningful experiences, such as appreciation of Da Vinci's use of *sfumato*. There is a striking similarity between this approach to thinking about objects and the geographer Yi-Fu Tuan's thoughts about the nature of place. Tuan conceived of space, unbounded and undifferentiated, as like movement, and place as akin to the pauses in the movement (which is not to say that places themselves may not be bustling with activity).[10] The metaphorical pause is what allows us to develop the kind of relationship that transforms space into a place that we are emotionally invested in. Both approaches, whether couched in terms of time or motion, present us with the idea that it is a pause in the rush that allows for meaning to accrue, the way that lichen collects on a stone in the river.

What do we learn by looking at a material object, something we are inclined to categorize as fixed in time and space, and recasting it as an event in the process of unfolding? Or, perhaps more aptly, a thread in the process of unspooling, even fraying? Events are transitory, and so we are more inclined to approach our experience of them in a manner that is sensitive to their passing nature. If you want to see the sunrise from the top of Cadillac Mountain (arguably the first place the sun hits on the Eastern seaboard of the United States), then you need to know when to be at the top of the mountain on a given day, wake up early enough so that you have time to get there, and so forth. Moreover, we expect that the event will have a limited duration. The sunrise doesn't take place all morning: there is a moment, imprecise though its arrival may be, when the sun is just *up*, and the sunrise is over. If you came to see the sunrise, that's the point when you're likely to head back down the mountain, maybe to take a nap or get another cup of coffee. All of this is straightforward, and we could tell similar stories about other kinds of events: attending concerts, going to the movies, participating in a ceremony, and so on. At some point they begin; at another they end.

Why do we expect that a work of architecture should always be available for our appreciation rather than viewing this assumption as presumptuous, like the expectation of a suspended sunrise? Why think that a painting should stay the same over time? By viewing objects on the model of events, we might recognize that there are certain times in the life of the object that will afford special kinds of meaning, distinctive kinds of experience, but those periods are not endless. For different materials, these moments might be quite long, so extended as to strain at the boundaries imposed by a word like "moment" that conjures a sense of brevity. We don't necessarily err in trying to extend the period in the life of an object that renders a certain meaning available to us, but we can also recognize the further, if different, value that can attend the moments after. And we do—we find it in our appreciation of things that are distressed, weathered, patinated, ruined.[11]

Dominguez Rubio is clear that his goal in likening objects to slow events is not to direct our attention to processes alone, but to think about the context of conditions that allows objects to have particular meanings at particular times, and hence what work might be required to maintain such conditions. As he puts it, we should think *ecologically* about conservation.[12] While the idea of ecology here might conjure images of environmentally friendly conservation practices, that's not the intention. The point is that you can't regard the object in isolation from its environment, the context in which it is valuable, or its relationships to us and other things. Or, rather, you *can*, but then you risk conserving only material—like a genetically authentic crow that doesn't know how to act like a crow—not the meaningful object itself. To use a term that sheds the more environmentalist connotations of ecology, we might call conservation that takes seriously the relationships among objects, people, and places a *situated conservation*.[13]

—

The influential scholar of cultural law and policy, John Henry Merryman, outlines three considerations that an object-oriented cultural property policy should take into account: preservation, truth, and access.[14] He identifies preservation as most fundamental, to be prioritized over truth, which he glosses as the knowledge we can gain from studying an object, with the value of scholarly and public access taking up the rear.[15] While Merryman acknowledges that these considerations are related in practice, he claims that they are "conceptually separate." What the foregoing discussion about a situated approach to conservation indicates, though, is that preservation, truth, and access are much more conceptually intertwined than Merryman's schema suggests. To attempt to preserve the object outside of the context that makes it meaningful to relevant communities would risk preserving only a hollowed-out husk of the thing we cared about conserving. This point is obvious in more personal contexts, though we tend to forget or ignore it once we enter the space of public museums. You can't preserve the meaning of a family heirloom *as* a family heirloom by selling it to someone better positioned to preserve its material integrity.

A related point can arise in museum contexts as well. Miriam Clavir's important book *Preserving What Is Valued* emerged from the tension she initially felt as a museum conservator fielding requests for objects in her care to be loaned back to First Nations peoples.[16] Would the objects be safe beyond the protective boundaries of the museum? One of the major lessons of the book is that attention to *how* First Nations people think about conservation, which differs from views that predominate in Western museums, can help to resolve the appearance of a dilemma—the wear and tear of use is not inconsistent with conservation if it is the object's *role* in community practices that you're aiming to conserve. Put in terms we explored in the previous chapter, requests for the loan or return of certain cultural objects can thus be seen as efforts to manage their change on terms that the cultural community can accept.

This point can be generalized to all manner of contexts. It's not clear that things can even retain their identity, remain what they are, so to speak, outside of the appropriate setting. We might say this about recent artifacts that have been rendered obsolete by technological advances. Think about a rotary phone, for example. We might put it on display in a museum of design or technology, but is it still really a phone in that setting? With no dial tone, no connection to a network of wires that might put a human voice on the other end of the receiver, the phone becomes a stand-in. It refers to something that was once a meaningful part of our lives, but this artifact isn't quite that thing. It has become a relic.[17]

If this is true of household objects, the point becomes all the more salient for things whose identity is even more bound up with their meaning—artworks, heirlooms, dances, idols.

The Cambodian dancer and choreographer Sophiline Cheam-Shapiro describes how she was asked to leave the Metropolitan Museum of Art while performing a danced prayer to a (likely looted) Cambodian deity on display. As she put it: "Whenever I visit museums around the world that house Khmer antiquities, I pray to the gods and ancestors that inhabit them. Sometimes I simply put my hands together and chant. Other times I move. This is my tradition. It is an essential part of my identity and my relationship to these objects."[18] What it is for an object to be a Khmer antiquity (as opposed to just a stone statue) is in part to be an object of worship, and what it is for Cheam-Shapiro to maintain her Cambodian identity (as she understands it) is in part to relate to such objects in appropriate ways. The philosopher Alva Noë, leveraging the role that continued performance must clearly play in the conservation of choreography, puts the general point about art conservation like this: "But isn't that what we need to do, too, if we wish to preserve paintings, sculptures, films or videos, architectures, or whatever? It isn't just that we need to resist the microbes and the fungi and so keep things from falling apart; we need to preserve the very environment, or our access to the very environment, in which alone the works are actually even the works they are in the first place."[19]

Consider these points with respect to contemporary street art and graffiti practice. Although works in this broad category are quite diverse, they seem to at least be united by a commitment to ephemerality—by placing them in the street, as opposed to a gallery, the artist recognizes that they will be subject to the forces of change that are typical of a street environment.[20] There are different rules and norms for different kinds of spaces, and the attempt to interpolate them can create deep tensions. Consider the building owner who tries to preserve a Banksy by covering it with plexiglass. In trying to impose museum norms of preservation on a piece of street art, he arguably kills it—it can't be the same kind of work anymore when covered in a protective coating, undermining its commitment to ephemerality. When street artists are invited to create works for gallery spaces, it seems they necessarily become engaged in a different sort of artistic practice, one that is divorced from some of the constraints and values that are definitive of street art and graffiti.[21] Of course, there are many examples of artists and provocateurs bringing the norms of the street in to challenge museum norms uninvited, including Banksy, the graffiti artist Katsu, and the comedian Tom Green (hello, children of the '90s), who each use different, non-destructive means to treat the staid halls of museums as canvases for artistic expression (or perhaps just a bit of fun in the latter case).[22] But again, in order for those works to function as intended, they need to be understood in relation to the norms of their environment—they couldn't be the same works outside of a museum or gallery.

We have already seen that a concern with maintaining identity can motivate conservation. A situated approach to conservation highlights how the work of conservation in turn shapes identity. By making choices about what is worth saving and why, we shift understandings of what we should retain and what we can let go—and that process itself influences who we are. The early 20th-century curator and art critic James Laver wrote: "Museums are graveyards, unless some central purpose, some controlling impulse can be found to give them life. Museums or mausoleums? That is the question."[23] A situated approach to conservation resists turning museums into mausoleums by attending to the context and connections that make objects meaningful. By broadening our perspective to include not just the object, but the system of relationships in which it is embedded, we can more easily see the pitfalls in focusing on the material alone, or the limits to the kind of conservation work that can be achieved when confined to the boxes where we keep those objects.[24]

None of these considerations are intended to imply that material conservation doesn't matter. Rather, they allow us to view material conservation in a context that is guided by other values beyond preservation as an end in itself. The extremely risk-averse person who is coaxed out of reclusion is not being told that that their physical or mental health isn't worth attending to—but, rather, that a life engaged with the world (and even some degree of risk) grants the preservation of health a different meaning, one that enables a broad range of valuable activities and experiences.[25]

This discussion has implications in turn for any conservation practices that end up treating the material approach as a paradigm. Consider the case of language preservation. How do you save something like a language? If you take material conservation as a paradigm, you might focus your attention on documenting and archiving the language. That work has its place, but it can distract attention from the value of conserving *communities of language speakers*. A recent academic project has pioneered a method for creating 3D-printed models of grammatical structure. These are fascinating objects and could have symbolic value for spurring action against linguicide, or of offering an intriguing intermodal perspective on language diversity. But they should also offer a lesson about the potential for misuse of material paradigms for conservation. It would be an error to think that these models themselves might constitute the preservation of a language. A news story about the method reports: "The first of its kind in the world, the project and its outputs not only enable the visual demonstration of the architecture of language but also enables its preservation in a permanent, solid form."[26] It's seductive to think that if we can somehow render the intangible in material form, we will have saved it for posterity, as if being able to hold it in our hands would ensure that it could never be taken away. But this is ultimately akin to using taxidermy as a method of species conservation. Yes, you will have conserved something that stands in an interesting relation to what you set out to save, but it won't even be the same *kind* of thing, let alone the thing itself.

While this case helps to highlight the pitfalls of material paradigms for conservation, it should not be taken as an indictment of digital methods as a useful conservation tool. As with any method, we need to see how it's being employed in order to assess whether it is effective. Pickling might not work well for preserving chocolate bars, but that's no indictment of the method's value for other foods. Many digital methods are criticized for being "inauthentic," but as with the discussion of crows that opened this chapter, when we broaden our perspective on authenticity, we may see more potential for digital interventions than initially meets the eye. We can see this by turning back to cases of replication (digital and otherwise) that we first introduced in Chapter 2.

If you visit the website for Google Arts & Culture's "Open Heritage" project, you will see the tagline: "Preserving our shared heritage." How is the Open Heritage project pursuing this preservation work? In large part, through 3D modeling. As with the case of the 3D-printed language model, there is something initially unnerving about the idea that scanning cultural sites that face threats from armed conflict or climate change is really preserving them in the sense that we care about. Yes, if the sites are destroyed, we will have another representation of them to add to photographs and videos, maybe even a representation that we can explore in virtual reality. But the sites themselves will still be gone. UNESCO helped create an educational program called History Blocks that allows students to recreate destroyed heritage sites in Minecraft. There is no doubt promise in connecting children with history through video games: I'm a fan! But we need to be clear about what work these digital interventions do and what aims they leave behind. I came across a 2019 article about the History Blocks project titled "No War Can Destroy These Minecraft-Made Heritage Sites."[27] That's a complex and puzzling statement. For one, it reflects the widespread misconception that digital media is somehow impervious to destruction and doesn't itself require constant maintenance to remain accessible. Ironically, when I clicked the article's link to historyblocks.com in early 2023, it was dead, a mere four years later. But furthermore, the statement, akin to the slogan of the Open Heritage project, risks inviting us to view digital representations as replacements for the sites that they model.

One risk of this implication is that it may invite us to adopt a more cavalier attitude toward the preservation of the sites themselves; after all, we can always recreate them in virtual reality.[28] While I would hesitate to attribute this explicit view to most practitioners in the field of digital conservation, a concern with such a consequence, even if unintended, is not unreasonable. For instance, in an interview with the BBC, archaeologist Bill Finlayson worries that "the dangerous precedent (of on-site reconstruction) suggests that if you destroy something, you can rebuild it and it has the same authenticity as the original."[29] When the Oxford-based Institute for Digital Archaeology created their 3D-printed model of Palmyra's arch of triumph, which was destroyed by ISIS, the *Guardian*'s art critic Jonathan Jones insisted that Palmyra "must not be turned into a fake replica of its former glory."[30] The specter of these concerns does demand that we ask: when we claim digital methods preserve our shared heritage, what exactly do we mean? How is digitization a method of conservation? These questions are especially pressing for those conservation candidates whose value is understood in terms of their embodiment of a particular history, and as we've seen in the previous chapter, most any conservation candidate can in fact be couched in such narrative terms. If digital methods don't themselves maintain that very embodied history, how can they claim to offer an avenue for engaging with what we valued about it in the first place?

The arch of triumph at Palmyra is an instructive case, because it is itself a ruin. One of the distinctive features of ruins is that they don't only embody the past, but they make that past manifest through their visceral signs of wear and tear, the perceptual features through which they earn the name *ruin*. But an emphasis on the appearance of age might lead us to forget that not every compelling representation of age actually embodies the history it refers to (recall the discussion of *age value* from Chapter 2). We are familiar with simulated ruins from contexts ranging from the follies of landscape architecture to the acid-stained collegiate gothic buildings of U.S. college campuses. A replica ruin might be very convincing, but its ability to simulate the look of age does not by that stroke render it old.[31] If we are interested in being in the presence of an object with a particular history, no facsimile will offer us that kind of access. This suggests that whatever else a replica might provide by way of facilitating our experience of a ruin (or other old object), at the very least it cannot function as a replacement in this regard.

But it is precisely by resisting the idea that replicas or digital representations are primarily replacements or stand-ins for originals that we can simultaneously avoid the aforementioned concerns and position ourselves to understand the value that digital methods can offer for facilitating conservation of the *role* that objects, practices, and places play in our lives.[32] In Chapter 2, we saw that attention to replicas and their purported inauthenticity can help reveal *what* we are really interested in saving when it comes to conservation; we saw that identity creates a framework for conservation work that informs, and is informed in turn, by how attributions of authenticity are made. We're now in a better position to see *how* conservation (including replication) can do this work.

Meaning is created and accessed through various modes of engagement. Yarrow interviews Clare, a conservation officer, who emphasizes the different tools she uses in her architectural conservation work: "What Clare sees in these buildings relates to *how* and *with what* she sees them."[33] These include written records, photographs, architectural plans, and maps, including those created through GIS software. These tools facilitate an understanding of an object such as a building as much more than just material, but as an embodiment of history, an understanding which in turn conditions what is meant by referring to its "character" or "authenticity." In other words, a diverse set of tools, both digital and analog, provide the context needed to identify a building as authentic in the first place. But these tools aren't merely evidential; they not only reveal but in some cases *create* the relationships and narratives that authenticity consists in. The point, then, is that ways of understanding and engaging with a building, including digital replication, shouldn't be viewed as immediately raising the specter of inauthenticity—they may in fact be central to a dynamic understanding of what it means for an object to be authentic in the first place.

An excellent illustration of this point is found in the ACCORD project (Archaeological Community Co-Production of Research Resources).[34] In this project, researchers worked with ten different community heritage groups in Scotland to co-design and co-produce digital (and sometimes physical) representations of local heritage sites. The elements of co-design and co-production allowed community members to direct the selection of sites and objects to reproduce on the basis of their local significance, and to be actively involved in the technical process of generating the representations. Based on qualitative studies conducted throughout the process, the researchers note that "The ACCORD project provides clear evidence that the digital models accrue similar values to originals, becoming imbued with symbolic associations relating to identity and place."[35] But in addition to developing their own dimensions of authenticity, the researchers found that the replicas also helped to inform and enhance interest in originals as well, bolstering the point that replicas are not helpfully construed as replacements.

Crucially, the link between the digital replicas and the attribution of authenticity was forged by the role of community participation in the production of the replicas. As one participant noted: "It's authentic because *we* made it."[36] Performing the work of creating digital replicas in turn created connections among people, objects, and places that allowed participants to become part of the story they were uncovering, another reminder of the idea that history continues to unfold and weave new threads into its tapestry. The evidence provided by the ACCORD project concerning the role of community participation in lending digital replications an aura of authenticity should come as no surprise given the theoretical framework for situated conservation we have been developing over the past few chapters. What makes the authenticity of these original objects and sites meaningful in the first place is the role that they play in the lives of a community. No wonder, then, that the participation of that community in the production of replicas grants those products a connected significance—community involvement in the replication process also replicates the context of meaning in which the originals are embedded.

Community participation also helps to highlight the specific ways that digitization can be involved in the democratization of conservation practice. Too often, it is assumed that the mere act of digitizing something and putting it online is an act of democratization because it offers a new avenue for access that requires only an internet connection. However, as some commentators have noted, the assumption that making something available online guarantees access is faulty: there are barriers to internet access itself, government and institutional regulation of content, and inequalities built into particular platforms (such as the language used), to name a few examples.[37]

Even digitization projects that at first glance might seem straightforward are shaped in important ways by context and participation. For example, you might reasonably assume that the digitization of text would be simple enough—just as many of us have transitioned to ebooks, digital access to ancient texts seems like it might offer a more convenient and universal means of access. However, as the historian Raha Rafii argues, the supposition that manuscript digitization is an unequivocal good is more complex than it may initially seem. Focusing on examples of pre-modern Arabic manuscripts, Rafii cautions that a "philological orientation . . . frames manuscripts as texts to be read, front to back, rather than contextualizing them as cultural objects."[38] Methods that succeed in treating manuscripts as cultural objects and not merely texts will thus require technological approaches that are far more complex than the kind of "copying" of text that is familiar to everyday readers from ebooks. Moreover, even in the case of manuscript digitization, the participation of relevant communities can shape the selection of content to be digitized, as well as the methods employed. As Rafii puts it: "The control over historical and cultural narratives is often determined by access to sources, and uneven access, particularly when predicated on minimal access or effective denial of access for origin communities on an international scale, reinforces this neo-imperialist dynamic."[39]

Although the case of digital replication is fairly specific, we can see these themes echoed in all manner of conservation contexts. Discussing Military-to-Wildlife Refuge restorations (former military lands that have been remediated to serve as local species habitats and sometimes recreation centers), Marion Hourdequin and David Havlick emphasize the importance of the social dimensions of restoration work. Drawing on the idea of "focal restoration projects" that "engage people in the process of restoration, promoting community involvement and democratic engagement in restoration decisions," they illustrate how participatory restoration practices can work to not only restore the landscape, but restore the relationship between people and place through their very involvement, in contrast with a technical, expert-driven restoration project that simply hands the finished product over to the community.[40] In an altogether different context, in discussing the conservation of practices surrounding the harvesting of wild rice for Anishinaabe peoples in the Great Lakes region, Kyle Whyte highlights the way that food can function as a "hub" for a whole network of social relations that are "biological, ecological, cultural, economic, political, and spiritual."[41] The key is that these goods are not secured by wild rice alone, but by the traditional social processes through which it is cultivated, harvested, and distributed. These participatory goods could not accompany, for example, the U.S. government distributing bags of rice to Anishinaabe peoples. The goods in question require community participation to be achieved.

From digital replication projects in Scotland, to digitization of premodern Arabic manuscripts, to Military-to-Wildlife restoration in Colorado, to wild rice harvesting among the Anishinaabe: we could add case after case to a list of conservation practices whose value and success have been described by scholars, practitioners, and citizens as rooted essentially in community engagement.[42] This is because the practices generate a particular kind of good, what some political philosophers call "participatory goods." As the philosopher Suzy Killmister puts it: "What the participatory model draws attention to is the way in which practices such as language, ritual, and history provide the context within which the self is understood, supplying the scaffolding upon which personal identity is built."[43] More even than scaffolding, which is removed when a building is complete, these participatory goods often end up functioning as constituent parts of personal identity. They're not just the scaffolding—they're the brick and the mortar.

What's distinctive about participatory goods is that their value is dependent on being shared: they are not goods that we can simultaneously pursue in isolation from one another, but are the result of activities that we participate in together. It's not incidental that the practices described earlier are *community* practices; they have an essentially social nature that provides the environment in which the relevant goods can even be realized. Some of these practices will be structured, and different community members may occupy particular roles in sustaining them. For other practices, the relevant role may simply be participation in the group. But in either case, these forms of social participation involve recognition by other group members—that we, together, are working toward a particular end. The earlier examples highlight the way that conservation practice can *itself* be a participatory good, not just because it generates a certain product, but because it can be a practice that engages a community in conservation of the very context of meaning in which that project is situated.

Although participatory goods in this sense stem from participation in a community, we can take a lesson from them that can apply in individual contexts as well. We might call this an *engaged* form of conservation, a subset of participatory conservation characterized by active involvement with a conservation process, but primarily as an individual rather than in collaboration with a community. Consider the difference between working to maintain a treasured object yourself versus completely outsourcing its maintenance to a third party. Outsourcing conservation might produce a maintained product, but it won't involve your actually engaging in the process of caring for the object yourself. Even individual conservation projects can be "focal" in the sense discussed earlier, then, when they involve acquiring knowledge, practicing skills, making choices, and so forth, which deepen and shape one's relationship with the conservation candidate. This is not to imply that no conservation work should be done by trained professionals—there will obviously be all kinds of highly skilled labor involved in successful conservation in contexts from works on paper to environmental remediation. But as we saw in the case of classic Mustang owners, opportunities for individuals to engage in the process of maintaining their possessions through learning, discussing, building understanding, and so on, even when they don't perform all the work themselves, can transform their conservation into a hub for a host of related activities that can in turn enhance the meaningfulness of the conservation candidate.[44] Cultivating an attachment through conservation is what makes the felt needs the attachment generates function not only as constraints on action, but as anchors for identity.[45]

We see this kind of engaged involvement in conservation methods in the ACCORD project. While experts facilitated the participants' use of technical digital tools (photogrammetry, reflectance transformation imaging), participants' direct involvement in the co-production of digital objects afforded the opportunity to engage in a web of practices and interactions that would not have been possible if that work had been completely outsourced to experts. This points in the direction of a growing concern we might have about the use of AI tools in digital aspects of conservation and restoration work. If such tools end up removing key aspects of human work and eliminating the elements of co-production that might otherwise be involved, then they will compromise the ability of the conservation *process* to be an integral aspect of the work that conservation does, and hence its overall value. Even when we're not approaching the use of these tools with an eye to conservation, the replacement of human creative activity with AI-driven products might by its very nature hold implications for conservation by inhibiting the continuation of particular practices and traditions. For example, imagine a basket weaving tradition that has been handed down through generations. If the production of such baskets could be outsourced to an AI-driven 3D printing program, it could disrupt the focal practices and web of related activities involved in the traditional practice itself.

An especially striking example of outsourced conservation comes from a recent project intended to preserve the Māori haka, a traditional dance, via robot performers. The leaders of the project explicitly conceive of it as a form of cultural preservation intended to ensure the maintenance of the practice "even when humans are not available to perform."[46] In a detailed study, the cognitive scientists McArthur Mingon and John Sutton explain that the very idea of robots preserving the haka misconstrues the nature of the tradition. Haka is a form of embodied knowledge that involves the complex expression of inwardly felt emotions in response to particular events and environments and in dialogue with the audience. It involves the preservation of knowledge, but not in an "archival" sense that could be simply recorded or programmed—a robot programmed to perform a dance that looks like a haka is simply going through the motions. As they put the point: "the practice and performance of haka *is* the preservation of Māori embodied knowledge in action and is a process that cannot simply be offloaded to robots."[47] It is precisely in failing to be participatory and situated that the robot Māori haka is ultimately incapable of preserving the haka at all. In other words, there will be cases where outsourcing conservation can't even ensure the successful preservation of a product, let alone a process, because it mistakes the relationship between the two.

This can be especially true when objects or places come to have an explicitly symbolic value. Yarrow notes that for Clare the conservationist, an approach that takes buildings as being symbols or emblems that embody particular moments in time "underscores a personal sense of ethical commitment,"[48] a desire to do what's "right" for the building under that emblematic description, even when this might be in conflict with the wishes of current owners. But sometimes it's the current occupants of buildings who are most committed to their emblematic value. Consider the story of the Brooklyn headquarters of the United Order of Tents, a secret society of Black woman with its roots in the antebellum period. Contemporary members of the group are working to maintain this historical home of its eastern district. "That building is a beacon of light," member Akosua Levin said. "It's a beacon of where we came from, and where we can be. We have a foundation. If we get weak: Look at that building. If we feel like we can't go on anymore: Look at that building."[49] But the ability of the building to function in this way is itself dependent on the participation of the Tents members in the community that makes the building meaningful. Without the ongoing activities of the organization, the building would still embody the past, but it would fail to be emblematic, especially if its story is forgotten. In this way, the work that the Tents members are doing to save and restore the building involves not just material and financial labor: it involves the work of maintaining the context in which the building remains meaningful, which in turn helps to strengthen the community itself.

Participatory and situated conservation work can allow us to cultivate *personal* relationships with the things that we take care of.[50] Outside the context of family heirlooms and keepsakes, it is all too common to find the value of conservation candidates couched in *impersonal* terms, especially among advocates of their universal value (a topic we'll return to in earnest in Chapter 7). They have a beauty, or history, or function that anyone might care about, regardless of who they are. As the philosopher K. Anthony Appiah presents the idea: "The connection people feel to cultural objects that are symbolically theirs, because they were produced from within a world of meaning created by their ancestors—the connection to art through identity—is powerful. It should be acknowledged. The cosmopolitan, though, wants to remind us of other connections. One connection—the one neglected in talk of cultural patrimony—is the connection not *through* identity but *despite* difference."[51]

But to say that anyone might value something need not require that they do so by abstracting away from the specifics of their own identity: "through identity" and "despite difference" need not be in opposition. Something may be worth valuing from within a wide variety of particular perspectives, so that we find that despite our many differences, there are points of access for appreciation that are couched within many radically different identities. In other words, we need not view things from the thin perspective of our mere humanity, as the cosmopolitan would have it, in order to find commonality that overcomes difference. That we are each multifaceted beings with a complex identity offers a richer form of commonality than appeal to our species membership.[52]

To say that we each experience the world of value filtered through the perspective of our own identity is not to suggest that such experiences must be private or exclusive—these perspectives themselves might be shared. For example, part of our conservation work might aim to cultivate particular moods or shared emotions at a site. We might think of a group emotional state as involving a mutual awareness of each other having simpatico emotional experiences.[53] For example, Tom Cochrane argues that when listening to music together, we aren't just experiencing the same emotions at the same time, a group of people having private individual feelings disconnected from one another, but are rather "using the music to mutually determine" our emotions.[54] It is this mutual recognition and shaping of feeling that makes the emotional experience qualify as *shared*, and just as music can endeavor to cultivate certain moods, so, too, can other conservation candidates. It is precisely by making the preservation of the Brooklyn headquarters of the United Order of Tents the focus of a shared conservation project that the Tents cultivate and maintain its status as a "beacon of light," a site that can foster a particular kind of emotional attachment for the group. The building can't do this on its own; it is a force that is nurtured through the act of caring for it.

Participatory and situated conservation can thus alter the terms on which we confront change. We saw in the previous chapter how the source of change can alter its valence: malicious destruction generates a different meaning from natural degradation. Sometimes our conservation efforts will target altering the source of change itself; to preserve a traditional practice threatened by oppression is not to insist that it never change, but to take a stand on who will determine the nature of the change it undergoes. But we cannot pursue this kind of conservation alone. It is by collaborating with others, exercising our agency to influence how the things that we value will undergo change, that we take control of the terms on which change is experienced, sometimes even in the face of overwhelming power. Even if these conservation efforts are ultimately unsuccessful, simply to resist them is to shift the meaning of the change, the same way we might differently assess a losing sports team that rolls over and gives up in contrast with the team that leaves it all on the field. To take up the mantle of conservation can itself be a way of shaping the terms on which change is experienced.

My claim is not that conservation that fails to be situated and participatory in the way I have described here is worthless, but that it lacks the promise of a richer meaning. The late philosopher Joseph Raz writes: "the point of values is realized when it is possible to appreciate them, and when it is possible to relate to objects of value in ways appropriate to their value. Absent that possibility, the objects may exist, and they may be of value, but there is not much point to that."[55] We may at times be able to preserve things in ways that disconnect them from the meaning that they hold for peoples' lives. Perhaps there will be occasions when that is a worthwhile enterprise; but as Raz observes, it's much harder to see the point.

6 FROM APPROPRIATION TO PARTICIPATION

BY WHOM SHOULD THINGS BE SAVED?

It is hard to imagine a preservation topic more sensitive than the protection of human burial sites, and were it not for the countless stories of grave desecration in the United States, it might be equally hard to imagine that such issues would even arise. Who, you might ask, if your family has not been the victim of such an affront, would destroy a grave? The U.S. government, for one. In 1867, an order from the Surgeon General of the United States directed members of the army to find and recover skulls and remains of Native Americans for research purposes.[1] You can add local governments to the list, not to mention private developers: many cemeteries that hold the remains of African Americans (whose burial grounds, too, were segregated during the antebellum period and beyond) are themselves now buried under highways and skyscrapers.[2]

Having seen the role that participatory conservation can play in the interwoven work of maintaining both conservation candidates and a sense of self, it should be clear that there's a lot at stake when it comes to who is engaged in conservation projects. If conservation is bound up with questions about individual and collective identity, then it is essential that the relevant parties have the agency to shape conservation choices—this is, in part, a way to shape how they see themselves. If you take things that are cherished by or even sacred to a people, such as burial grounds, and wrest those sites completely from the control of families and communities, in doing so you assert a perverse authority over them. There is a trope in books and movies where the villain gets his hands on some fragile and essential object, then carelessly crushes it in his hands or under his heel, while the protagonist looks on in gutted horror. What makes these scenes so wrenching despite being commonplace is the way they embody a phenomenon that is all too familiar—someone else exerting complete control, even to the point of destruction, over the things we care about the most. Examining the effects that *lack* of influence and control in conservation projects can yield will help us bring into even sharper focus why participation in preservation work can be so important.

Precisely because of the relationship between conservation candidates and identity, the objects, practices, and places that we care about can come not only to be core to our sense of self but to *represent* us as well. It is because of the role that a local sports team plays in the life of a community that locals can see the team as representing the town or city; it is because of the importance of certain cultural practices, whether dances or cuisines, that they are adopted by the people who hold them dear as representative of their culture. Of course, things can also come to represent groups in ways that the community itself would reject—these associations may ultimately end up functioning as caricatures or *mis*representations. Sometimes these misrepresentations are a kind of illicit synecdoche—as when an incredibly diverse category of people comes to be symbolized in the dominant imagination by cultural features specific to only a subset (some generic racial stereotypes would illustrate the point). When those insufficiently familiar with the details of a community's life exercise authority over how their values are presented, misrepresentation is likely to occur.

The risk of misrepresenting people through things (objects, practices, places) is among the worries that can motivate accusations of *appropriation*. Moral and political objections to appropriation can arise in all manner of contexts, from food to fashion, music to museums. The kinds of concerns that objections to appropriation raise are as varied as these settings. To the extent that there is a common thread running through objections to appropriation, it is that these concerns surround members of a dominant group taking, using, or controlling things that belong to, stem from, or are otherwise associated with a marginalized group. The generous helping of descriptors in the previous sentence gestures at the extremely wide range of ways that objections to appropriation are made, and the consequent difficulty (perhaps even impossibility) of helpfully elucidating the problem as a whole.[3] Luckily, we can restrict our discussion of appropriation worries here to the ways that they arise specifically in the context of preservation and conservation; but the issue remains complex.

Debates about appropriation are often portrayed unhelpfully as a conflict between two diametrically opposed positions. On the one hand, you have those who rally around the banner of freedom, and insist that anyone anywhere must be able to create, own, and profit from anything they fancy (within the bounds of the law, perhaps), free from moralizing restrictions that would place limits on their creative or entrepreneurial liberty. On the other hand, you have those who would insist on the precise division of the world into tiny bounded wholes exercising complete control over their own cultural products as a safeguard against the furtherance of oppression. These descriptions are both caricatures, to some extent, though representatives of each view do crop up in public discourse on appropriation.

As is often the case in disputes between starkly opposing sides, both camps get something right, though neither view is satisfactory as it stands. The first camp is correct that freedom is a crucial value, but doesn't adequately recognize that freedom to act should not entail moral license to simply ignore the consequences of those actions, and certainly doesn't include freedom from being confronted with objections to those actions. The second camp is correct that systemic injustices have created a cultural landscape that is far from level and that this warrants moral care in the navigation of cultural space, but it doesn't adequately recognize that overly restricting the space of cultural experimentation and exchange is both morally undesirable and conceptually untenable. The tricky task is to figure out how to reconcile the desirable elements of the two positions while avoiding their significant pitfalls.

It may be helpful to think about these issues in a fictitious individual context before scaling up to the realm of groups. Imagine that you live in a small town, and your grandfather played a key role in the town's founding, helping to construct its first buildings. You still own the red wheelbarrow your grandfather used (so much depended upon it!), and you like to keep it leaning against the porch as a personal homage to his part in town history; the working conditions were not outstanding, but you're proud of what he accomplished. Now imagine that one day you leave the house to find that the wheelbarrow is gone, because I, a resident of a more affluent town nearby, have taken it. I've put it in my garage (you don't have one) where it will be safe from the elements, and I plan to create some didactic materials that will help educate people about your grandfather's life and importance to the town; in particular, the storied day that he flipped the wheelbarrow over and stood on top of it to address the assembled townsfolk and quell a labor strike. You've never heard that story before, and it seems out of character based on what you know about your grandfather, but I'm quite the history buff, and I've done some research on this. Maybe I'll even charge some modest fee for people to see the wheelbarrow and learn about the town's founding.

Now, this may seem like a ridiculous tale, but many of the elements that it captures are eerily similar to real stories in the history of preservation work that are in fact far less benign than the antics in this story. In the real cases, items have often been stolen from graves and sacred sites, or acquired under conditions of colonial duress.[4] And it's not as if the *theft* of the wheelbarrow in my story bears the full weight of the moral concerns the tale evokes. If you like, you can imagine that, under difficult financial conditions, you've pawned the wheelbarrow and I have legally acquired it. The story is still likely to raise your hackles. It certainly seems that I've wronged you in this tale. How should we characterize those wrongs? It's not clear that history is being accurately or sensitively portrayed here, you won't reap any of the financial benefits of the fee I plan to charge, and you haven't been consulted or involved in my display in any way. These are worries about *representation*, *exploitation*, and *lack of meaningful participation*.

While the story I told focused on two individuals (you and me), when we begin to think about the impacts of these three moral concerns, broader groups of relevant parties quickly begin to take form. Worries about misrepresentation of your grandfather's life aren't only relevant to you, or even your immediate family—they have a bearing on the whole town and anyone with an interest in developing an accurate understanding of its history. There's a direct sense in which I would be exploiting you and your family in this tale, but possibly other laborers involved in the founding of the town, as well. And meaningful participation in the care and presentation of artifacts important to the town would plausibly involve others as well, provided that you were inclined to donate your property in that way. In thinking about scaling these appropriation worries from individuals to groups, notice that we're not only picking out preexisting groups and asking if they might be impacted. Some of those impacted might be part of such groups, and impacted specifically in their role as members of a group (your family, for instance), but others might not have any formal or social association with one another.[5] In a sense, then, we are also *constituting* groups by thinking through the impacts of a purported act of appropriation. The way that groups can be formed *through* our approach to conservation (or through the impacts of a failure to pursue it well) will be important; but first it will be helpful to get a clearer sense of how appropriation intersects with the problems of misrepresentation, exploitation, and participation.

—

Appropriation does not guarantee that preservation practices will misrepresent culture, but history suggests that the risks of misrepresentation are increased when outsiders to a cultural context are in charge of caring for the valued items of a community they don't belong to.[6] We have already seen one example of this in the case of the Zuni *Ahayu:da* sculptures. The control of these sculptures by Western museums not only ensures that they are not treated the way the Zuni intend (slow decomposition through exposure to natural elements) but, in turn, represents them to the public inaccurately by that very stroke. These sculptures are importantly unlike sculptures intended to outlast the elements, and so it seems to display them in a museum cannot *but* misrepresent their nature, at least in this regard. Zuni Pueblo has sought the return of *Ahayu:da* for decades, in many cases successfully, and one important aspect of understanding the value of these efforts (in addition to redress for the illicit acquisition of the sculptures) is to recognize the important role that Zuni people play in shaping accurate representation.[7] If we care about preservation of the sculptures as they are meant to be, rather than on the basis of a Western conception imposed upon them, then we should be invested in who is in control of their care. Accurate representation may not always require the leadership of an insider to the culture in question, but it will always require the knowledge necessary to avoid misrepresentation—and sometimes that knowledge is only available to group members, either because it is private or because it is dependent on experience that being a group member involves.[8]

Exploitation is a contested concept, but it can be helpfully thought of in terms of extracting excessive benefits from people whose ability to refuse has been compromised.[9] For example, consider aspects of contemporary "fortress conservation" in Africa, where local Indigenous communities have been displaced from their traditional homelands and prevented from engaging in subsistence hunting, while private companies that manage these lands, often led by wealthy Westerners, profit from tourist revenue.[10] This model echoes similar displacement of Indigenous populations in the creation of the U.S. National Park system in the late 19th century; again, all in the name of environmental conservation.[11] The appropriation of this land for preservationist ends is subject to multiple moral objections, but exploitation is surely part of that story.

Lack of meaningful participation from relevant communities is a feature that can be found in many preservation contexts where concerns about appropriation have been raised. It is worth stressing that *mere* participation is often inadequate to allay concerns about appropriation; as the curator Joe Horse Capture points out in an essay on the traveling exhibition "The Plains Indian: Artists of Earth and Sky," there is an important distinction between *consultation* and *partnership.*[12] Merely asking someone for input does not guarantee that they have any say in what happens. As Patricia Marroquin Norby, curator of Native American Art at the Metropolitan Museum of Art, explains: "Some tribes seek repatriation, while others favor a co-stewardship approach or prefer that works remain at the museum. Community needs are diverse, yet very specific. One commonality across communities and cultures is *the desire for a say* in how and if works are publicly presented, and how they are cared for."[13] Participation holds the promise of allaying the problems of both misrepresentation and exploitation, but only if that participation meets certain requirements.

We have already seen in past chapters that the terms on which change is confronted can shape the character of that change, in particular its valence and moral contours. The distribution of power and authority within contemporary institutions devoted to preservation and conservation makes the negotiation of these terms complex and morally fraught. It would be a mistake to think that merely involving some members of particular communities in preservation work is sufficient to alter the terms on which change is confronted. How precisely communities come to be engaged in preservation work itself contributes to the nature of these terms.

For example, the philosopher Shelbi Nahwilet Meissner explores the complex ways in which Indigenous participation in efforts to preserve both language and the environment can be negatively conditioned by the terms on which this participation is elicited and allowed by those in positions of institutional power. In the case of language preservation, Meissner argues that language revitalization efforts led by organizations such as UNESCO and U.S. and Canadian universities reinforce *dependency* relationships between Indigenous groups and institutions that are part of the very colonial systems that led to linguicide in the first place. They do this in part by requiring Indigenous communities to perform a certain kind of trauma, in particular portraying themselves as helpless victims in need of aid, in order to be eligible for access to resources. This, as Meissner puts it, "return[s] us to a narrative in which Indigenous people cannot heal from our imposed trauma without the assistance of the researchers, universities, and nation states complicit in our colonization."[14] Moreover, when language archives are maintained by these institutions, they typically end up *controlling* the archives, generating (false) dilemmas about how to navigate access to sometimes sensitive materials (by making the archives available to all vs. restricting access). What makes such dilemmas false is that they are ultimately a consequence of non-Indigenous institutions exercising control over archival access in the first place.[15]

In the case of environmental conservation, Meissner explains how Indigenous lifeways and practices surrounding relationships to the environment, such as through water, "were conceptually carved up . . . and rendered as parcels or mere resources" by non-Indigenous institutions.[16] For example, the process through which the U.S. government requires federally recognized Tribes to "quantify and monetize" water imposes a colonial way of thinking on Indigenous relationships to water: Indigenous practices and relationships go into the colonial policy machine, and "rights" and "resources" come out the other side. It may be that such resources are ultimately conserved by such policies and create some benefits for the relevant Tribes, but at the cost of forcing those communities to reconceive of their relationship to the environment on imposed colonial terms.

Both of these cases are incredibly complex and require the deep expertise of impacted parties to craft viable solutions. And surely the implication is not that major educational institutions or governments should forgo applying their considerable resources to support Indigenous peoples and cultures. My aim in raising these cases here is to highlight the way that preservation efforts that seem good on their face (Saving languages! Saving water access!) can be morally compromised by the processes involved and the terms on which relevant communities are able to participate in these projects. Just as forms of participatory conservation can generate participatory goods for communities, conservation practices that preclude participation on the right terms not only deny the possibility of creating such goods but can moreover be actively harmful to the communities whose identities are most closely bound up with certain conservation candidates. A conservation project that preserves a language by requiring Indigenous communities to relive and perform trauma, while fostering dependency on non-Indigenous institutions, is far from an optimal outcome, to put it mildly. Under such conditions, the relevant linguistic communities might aim to save their language as much from these non-Indigenous institutions as from "extinction."

It's worth attending to a subtle mechanism that helps connect conservation practice to the power dynamics discussed so far. The philosopher Sherri Irvin discusses the way that either applying or withholding conservation practices can facilitate an artwork's ability to express something through *exemplification*.[17] To exemplify a feature is to possess it *and* refer to it. As philosopher Catherine Elgin puts the point: "An exemplar . . . is a telling instance of the features it exemplifies. It presents those features in a context contrived to render them salient."[18] Fabric swatches and carpet samples are classic examples. A carpet sample possesses the features that a potential carpet will (its color, texture, etc.), and it also is designed to refer to those very features—it presents them to the consumer in order to facilitate making a choice about the features of the carpet. Importantly, though, the context helps determine *which* features are salient. For example, the size and shape of a carpet sample are not being exemplified (unless perhaps you're looking to install a carpet in a doll house).[19]

On Irvin's view, conservation (or the withholding of conservation) can be a way of manipulating the context in order to maintain certain meaningful features, not unlike Dominguez Rubio's discussion of how conservation endeavors to maintain the state of a thing that allows us to experience it as an object—this can be part of the process through which exemplars can be especially *expressive*.[20] For example, Irvin explains how the expressive power of Kara Walker's *A Subtlety or the Marvelous Sugar Baby, an Homage to the unpaid and overworked Artisans who have refined our Sweet tastes from the cane fields to the Kitchens of the New World on the Occasion of the demolition of the Domino Sugar Refining Plant*, a monumental sugar sculpture, was heightened through its exemplification of the process of dissolving: "the literal exemplification of impending destruction [. . .] activates a metaphor of the fragility and preciousness of our beings, the power we have over each other, the tragedy of disrespect and disregard."[21] Not only is the sculpture dissolving but it is referring to the process of dissolving in order to deliver a point: it is expression through exemplification. On the flip side, Irvin describes how Marc Quinn's self-portrait busts, made from his own frozen blood, express the visceral "precarity of human life" precisely *through* the intensive maintenance required to keep the blood in its frozen state.[22] To maintain Quinn's work or not maintain Walker's, then, is part of how the context is contrived to make features salient, fostering their expressive power.

But conservation doesn't simply function as an external force, a kind of scaffolding that can facilitate the expressive power of an object while remaining outside of it; by manipulating the context of maintenance, display, and reception, conservation becomes part of how we understand the conservation candidate itself. Conservation doesn't just help candidates to exemplify, but conservation practice *itself* can exemplify, embedding what it both instantiates and refers to in our understanding of the conservation candidate. Irvin is explicit about this elsewhere, explaining how the rule that an artist defines for conserving their work is "part of the artist's artmaking activity, and the rule is part of the structure to which it is appropriate to attribute meaning."[23] Quinn's self-portraits can't even be what they are without the conservation apparatus designed to maintain them—it is as much part of the life-blood of the artworks as his own blood. But the power of conservation to exemplify extends far beyond the world of contemporary art.

Many practices can bear the hallmarks of exemplification, whether literal or metaphorical. A ceremony can both confer honors and refer to them by putting the recipients on a pedestal; a public apology involves seeking forgiveness, but also makes that bid especially salient, exemplifying contrition. Appropriative practices, too, can function as a particular kind of exemplar. Even when an apparent act of appropriation isn't best understood in terms of misrepresentation or exploitation, it can generate moral concerns when it ends up *exemplifying* oppressive power dynamics through lack of participation.[24] Imagine a version of the exhibition that Horse Capture criticized that avoids any misrepresentation and provides the kind of access to the relevant communities that he recommends, assuaging worries about exploitation. Nevertheless, the lack of adequate participation from Native curators and experts *makes salient* who exercises power and authority in this context. It instantiates the marginalization of Native experts through their exclusion from the exhibition, a feature made especially striking given the exhibition's subject matter.

This illustrates why lack of meaningful participation ultimately sits at the heart of worries about appropriation in the context of conservation. Anyone can (though not everyone is likely to) misrepresent or exploit, but lack of meaningful participation by relevant parties in preservation can have an expressive power that is impossible to avoid when differential power resulting from marginalization is at play. Outsiders whose authority gives them the means and access to engage in preservation work, however well-intentioned, without meaningfully including the participation of insiders to the objects, practices, and places they aim to conserve, double down on their excess of power, and end up making a point of it through this very act of exclusion. In the case of museums whose collections have been built through colonialization, exclusionary conservation practices exemplify the very exercise of power that led to the acquisition of conservation candidates in the first place. We can tell a parallel story about environmental conservation in colonized countries, or even about the way similar power inequalities can shape exclusion from conservation in familial or interpersonal contexts.

But while conservation can seem like the culprit in these cases, it can also be part of the solution. Just as exclusionary conservation practices can serve to exemplify relationships of oppression and marginalization surrounding certain goods, so, too, could participatory conservation practices alter the terms on which change is confronted by exemplifying cultural resilience and autonomy. Recall Chike Jeffers's point that preserving cultural traditions can serve as an act of resistance against oppression. In order to achieve this goal, the oppressed parties need to play a role in preservation. If institutions that take the lead in the conservation of an object, practice, or place are bound up with the very forces that threatened it in the first place (as in Meissner's examples), then their work, even if it is successful in saving the conservation candidate, would exemplify the same assertion of power that created the problem—it would express the dependence of the marginalized on the powerful. It is only by actually *having the features* of collaboration and participation that a conservation practice can exemplify resilience and resistance.

Just as the manner in which Quinn's self-portraits are maintained becomes a part of how the artworks are situated and understood, so, too, can participatory conservation practices become part of how we experience and appreciate the things that they preserve. This is most apparent when we ourselves are involved in conservation. You may benefit if others in your community successfully protect a span of open space from residential development in your town, but if you were involved in the preservation campaign yourself, the saving of that land promises to mean something different to you. This is because you've changed what it exemplifies—by working to protect the land, its preservation is, in some part, the result of your action, and hence an expression of your commitment to the preservation of open space. You can't make that claim if you sat the campaign out, however much you may care about the issue. You can celebrate the campaign's success, to be sure, but it can't function as an *expression* of *your* commitment unless you play a role. And we can appreciate these shifts in expressive power as outsiders to conservation work, too. We can readily see the difference in meaning when a marginalized community is an equal partner in the preservation of objects, practices, and places that are central to community identity. We can appreciate this aspect of what is preserved, even when we're just empathetic onlookers rather than members of the affected community itself.

It's important to see that the goods of participatory conservation that I've outlined here are not achieved through mere representation alone. It is by all means a good thing for conservation organizations to hire and retain a diverse staff, for example, but the representation of people from marginalized communities in conservation organizations doesn't on its own achieve the kind of participatory conservation I've described. Participatory conservation isn't just about representation within existing organizational structures—it's about what those organizations *do* and how they engage with communities to further and shape their work.[25]

This point is crucial. Sometimes commentators think that worries about appropriation are allayed when they have *permission* from insiders to a group, but in a way that results in a kind of tokenism. As ever, context is key in thinking about the role that permission plays when it comes to appropriation worries. The kinds of groups typically implicated in appropriation concerns are unstructured social groups: it's unclear what the mechanism would be to receive permission that somehow spoke for the group. But in many practical contexts, permission on behalf of a whole group isn't what is needed, even where permission still seems to carry moral weight. Being invited as a cultural outsider to wear traditional clothing to a wedding, for example, offers an important moral license to do so—it's a context where wearing such clothing in the absence of an invitation would be puzzling, if not offensive. But this doesn't require any kind of cultural *group* permission, just permission from the hosts, and it also isn't a *blanket* permission to wear these clothes in any context.[26]

While some conservation contexts may create special circumstances where group permission is important, we should typically not be looking for permission on behalf of a group as much as we should be aiming for the achievement of participatory goods in our conservation practices. It's beneficial to achieve representation, but not even a member of your community can achieve the benefits of participatory goods on your behalf—if you want to glean the participatory goods, you need to find a way to get involved. This kind of participation can *result* in a form of tacit permission, which may create the impression that permission is morally relevant in such cases, but it's not what's doing the moral work.

We can now see that when it comes to concerns about misrepresentation, exploitation, and lack of participation in conservation work, the concept of "appropriation" won't always be central to our moral analysis of the case, even though it will often provide part of the setting in which criticism is made. This is because each of the wrongs we've outlined can occur even when appropriation by outsiders isn't part of the story. If we think back to our little fable about my theft of your grandfather's wheelbarrow, we can see that if we tweak the story so that I am your brother rather than an outsider, very little about the moral contours of the tale changes (in fact, some aspects may get worse). I've still misrepresented the story of our grandfather, I've still exploited you by not sharing any of the benefits of my project, and I've still excluded you from participation in shaping it. That I am an insider to the community in this case may give me a different moral claim to be involved in maintaining our grandfather's legacy, but it also doesn't absolve me from criticism for the various missteps I have made in the tale.

In these two versions of the story, we're confronting situations in which it is first clear that I am an outsider to the community, and then clear that I am an insider. But things are often much more complicated. One of the problems we face in thinking about appropriation worries at the group scale is determining who rightly belongs to a particular group. If the fundamental feature of appropriation involves the use or control of objects, practices, and places *across* group boundaries, then it seems we need to have some sense of where those borders are drawn.[27] However, unlike states, nations, Tribes, or other formally organized groups, the kinds of groups involved in disputes over appropriation are typically unstructured, amorphous, overlapping, and dynamic. To be sure, disputes about membership arise in the context of organized groups as well, but they pale in comparison with the complications we face in the case of unstructured groups.

The language of "appropriation" not only implies some demonstrable differentiation between groups but also makes tacit reference to an understanding of ownership or property. To appropriate something is not only to take it from another group; it's to take something that is *theirs*. For this reason, we commonly see the invocation of property claims (or a lack thereof) playing a role in analyses of appropriation issues. In some disputes over preservation of tangible cultural property, there will be clear claims of title to be made, and hence relatively straightforward application of legal frameworks. In other cases, especially in disagreements over ownership of intangible cultural property, matters become decidedly murkier. Without dismissing the importance of property issues in certain cases, it's important to emphasize that the moral concerns that orbit appropriation are not exhausted by issues surrounding property.[28] People can be impacted by misrepresentation, exploitation, and lack of participation even when we grant that they have no property claims to the conservation candidates in question. This is another reason why thinking about who is impacted by such preservation practices is so important—the shockwave created by conservation practices that run afoul of these moral problems will be broader than whatever property claims may be found at the center of the issue. Imagine that we are siblings, and our late mother has left a cherished family heirloom to you in her will. Let's posit that the heirloom is indisputably your property. That doesn't mean that you would be immune from moral criticism if you immediately sold it, or set it on fire for fun.[29] Property claims are important, but they're not all-important.

These worries about group boundaries and property claims don't mean that the appropriation label is never relevant. There are plenty of cases where people *self-identify* as outsiders to a community and where there are clear ownership claims at issue, and so the shoe might well fit. But as we've seen, whether appropriation, even in a clear-cut case, raises moral concerns is complex—it doesn't simply follow from the fact that there's an outsider involved in the project. We should be attending to worries about misrepresentation, exploitation, and lack of participation whether it's clear that appropriation is occurring or not. If there's confusion about whether appropriation is the right label, we can sometimes just drop our concern with that question and focus directly on these other important dimensions of moral analysis, which are the features that explain what can make appropriation wrong in the first place anyway.

Rather than starting, then, with the often challenging, and sometimes incoherent, task of trying to determine the one group who ought to be in control of conservation projects, we could instead begin by thinking through who might plausibly be invested in the conservation of the candidate in question, and inviting participation on that basis. This approach allows us to bring another dimension to our situated approach to conservation. Different conservation candidates will be radically diverse in how they relate to and impact different kinds of communities, and it is only by looking at a particular candidate in its context that we'll be able to think through who might have an interest in being involved in its conservation. Sometimes the answer will have to do with particular religious or ethnic groups. Sometimes it will have to do with involving people who live locally in contrast with federal actors. Sometimes it might have to do with other forms of community that cut through and across these groups: gender identity, sexual orientation, ability, and so forth. But in figuring out who might have an interest in being involved, we will often face a dilemma. Those currently in a position of authority over a conservation candidate might not be in a good position to *know* who could have an interest in getting involved with a conservation effort. To be sure, sometimes they have a good sense and simply can't be bothered. But in cases where those in positions of power might face epistemic limitations (that is, constraints on what they're in a position to know), they may engage in good-faith exclusion of relevant parties.

To address this kind of hurdle, it behooves conservation projects to take as *open* an approach to their work as possible, engaging in outreach that provides opportunities for relevant communities to make their interest known. Importantly, this might require more than just publicizing conservation projects and hoping that people reach out—it can also involve meeting people where they are, coming to them, *seeking out* their feedback and involvement.[30]

One of the benefits of a participatory approach to conservation, as noted earlier, is that it can *create* groups that are otherwise heterogeneous but are united around a joint commitment to caring for certain conservation candidates.[31] The members of such groups might not share the same beliefs or belong to the same religious or ethnic groups—they might not otherwise belong to the same social group at all. However, through their collaboration they can foster conservation projects that maintain goods that play a role in a diverse range of individual and group identities. They can come to form a new association united around conservation.

A nice example can be found in the award-winning work of Mary Evans, a young conservationist who coordinated a volunteer conservation program at the Spalding Gentlemen's Society in Spalding, England (the society is no longer restricted to men, the name notwithstanding). The volunteers aged in range from sixteen to ninety, and brought a variety of skills and backgrounds to their work. Evans recounts the benefits that have stemmed from coordinating the group, ranging from the problem-solving skills of the volunteers, to the increase in attention to the Society's collections through community involvement. The case illustrates how a heterogenous group of individuals can come to form an association, making connections that might otherwise have been unlikely, through participatory conservation work, generating positive outcomes for both the community and the collection.[32]

When communities are excluded from a role in making preservation decisions, sometimes they will take action to remedy the situation. We have seen a telling example in activism surrounding monuments to the Confederacy in the United States.[33] One thing that activist interventions do is offer a new route to community participation in the process of determining the monument's future, whether that activism involves protest gatherings, the construction of counter-monuments, or even vandalism. My aim here is not necessarily to justify vandalism (though I think there are cases when it can be justified), but rather to offer an explanation of what such interventions achieve.[34] They are a way of *asserting* community participation in the face of exclusion from the conversation.

We can see further illustration of these features in another contemporary dispute about appropriation in conservation: that is, debates about the repatriation of artifacts from Western museums. Advocates of institutional retention have sometimes claimed that they are best positioned to care for precious objects. Putting aside the fact that this claim is sometimes dubious, if not racist, we could grant a qualified version of it for the sake of argument without it undermining claims for repatriation. If we think about conservation in situated terms, it will sometimes be the case that allowing relevant source communities to care for an object is an aspect of what *successful* conservation itself involves, apart from a concern with the integrity of material objects (just as we saw in previous chapters). Sometimes this will be true when the relevant communities are still extant. But even when it comes to ancient objects, there may still be regional or geographic grounds to pursue particular community-based conservation efforts because it allows the objects in question to become part of the source region's culture again. Interestingly, advocates of retention policies grant the premise of this argument. For example, opponents of returning the Parthenon marbles to Greece have argued that they have become key aspects of British cultural identity, an argument that we see articulated in other "Universal Museum" contexts as well.[35] If opponents of repatriation grant that this mechanism is possible, that caring for something can shape its role as part of one's culture, then why think that process is best facilitated for the descendants of those who acquired the objects through questionable means, if not outright theft (depending on the case), and not by those who share some kind of regional or historical lineage with the place the objects come from? While there are real and important questions about why a regional lineage in the face of cultural discontinuity is important, the claim is certainly no harder to press for source regions than it is for the homes of encyclopedic museums. If museums want to insist that a history of caring for collections makes them part of the cultural identity of the museum's home nation, then they cannot avoid the admission that they are *depriving* source nations of cultivating just such a relationship.

There are lessons here that extend beyond the reach of museums. Museums offer an example of a centralized conservation model—you take a large collection of perishable materials and maintain them under one roof, where conditions can be consistently and carefully controlled. Compare this with a more distributed model, where conservation candidates are cared for by a wide range of actors who have the prerogative to treat them how they see fit under a broad array of circumstances (in small regional museums and cultural centers, in local schools and libraries, in their homes, etc.). Centralized conservation can offer some important advantages over distributed conservation, which can include (though does not guarantee) the application of special expertise, resources, and preferable conditions for preservation. But centralized conservation comes at the cost of compromised control.

Consider, for example, how the rise of streaming platforms has created a profound shift in our relationship to all kinds of artistic content, from music, to movies, to books. Rather than owning these materials, we essentially lease access to them from the streaming provider. From a conservation perspective, one way of looking at this situation is in terms of a shift from a distributed to a centralized model. You might irreparably scratch a DVD or misplace a book, whereas in the cloud, those works are safe and secure, available for you to access at any time.

But while the centralized model at least appears to offer improvements in reliable preservation with one hand, it wrests away control over that content with the other. DVDs might be more fragile in some respects than remotely accessed digital content, but you exercise a different kind of control over them. They can't be remotely manipulated, altered, censored, or removed. And this is precisely what we've seen happen with streaming content in recent years. From the removal of content at HBO and Netflix, to the remote censoring of children's books, the preservation benefits of centralized streaming models have come with a lack of control over that content that is completely foreign to a more distributed model.

The lack of control that we experience with respect to streaming content is a breach of one of the few domains where people can expect to exert any control over the things that they care about. As preservationist Ned Kaufman observes, for most people, the boundary of control over preservation ends at their own dwelling, if they even exercise it there. "As for the rest of the places which we love and on which our lives depend, we can do little more than hope they will be there tomorrow and the next day."[36] We are often completely powerless to protect the things we care about.

The solution to this problem is not necessarily to try to increase the amount of stuff that you own so that you can exert more control over it. After all, private ownership of things that many people rely on and care about is part of what generates the problem in the first place. A helpful concept that often arises in the context of conservation is that of *stewardship*. The idea of stewardship allows us to dissociate the care of something from ownership of it—a steward does not typically own what they care for, but they are responsible for it. So, we might steward things on behalf of people around the world, or steward things on behalf of future generations, without making any claim that these things are fundamentally *ours*. While stewardship can still raise the specter of imposed values, and can still be a front for paternalism about the care of object, practices, and places, even when vacated of particular claims attached to ownership,[37] it does offer a category of caretaking that more naturally separates the *by whom?* question from the *for whom?* question that we'll focus on in the next chapter. When we are preoccupied with caring for what we own, there is an implicit (though, of course, not necessary) sense that we are conserving it for our own sake, and perhaps for those who are near and dear to us. But when we steward something, the weight of that implicit assumption is flipped—we are more naturally viewed as maintaining it for the sake of others, even if we also benefit from and are shaped by that work. The idea of stewardship opens the door to conservation work that we do *together*, as opposed to promoting a focus on just saving what is ours.

How we engage in collective stewardship of the things we care about will inevitably be fraught. We won't always agree. It will require compromise and negotiation. But these difficulties for participation in conservation are part of what it means to be involved with others in a collective project—you can't get the participatory goods without them. Moreover, as Kaufman notes, people currently exercise so little control over saving the things they care about; the familiar difficulties of working together with others should be welcome in a context where we so rarely have the opportunity to exert any influence at all.

So, what kind of terms should we aim for when it comes to stewardship? It can't be the kind of absolute freedom that we exercise over what we own, freedom without constraints, because we often don't own what we steward, and we will need to pursue stewardship collectively with others. The kinds of terms that different parties can expect will need to be relative to the particular context of conservation: what makes sense for a family caring for an heirloom will inevitably be different from what will be appropriate for the conservation of a town park, or an endangered species, or an artifact acquired through colonial procurement. But we can at least roughly characterize the constraints and goals of an approach to stewardship and conservation that seeks morally acceptable terms. On the constraints side, we should seek terms that avoid *domination*, where conservation (or lack thereof) is subject to the complete authority of the few to the exclusion of other participants, a problem we saw exemplified in the destruction of burial sites at the outset of this chapter. On the goals side, we should seek terms that allow for the achievement of the kinds of participatory goods that we've explored throughout this book, terms that allow us to exercise our agency in caring for the things that help make us who we are. As we have seen, this will often require more than mere presence or representation, but rather, the ability of participants to bring their knowledge and values to bear on the conservation project.

The burden of seeking improved terms for participation in conservation isn't completely borne by those in positions of institutional power, though. There is a role for all of us to play. How might we take up the mantle of stewardship in our everyday lives? How might we be *more conservative* in how we approach our interactions with the world? By this, I, of course, don't mean running around trying to arrest change wherever we find it, but rather, being more engaged in processes of determining what to save and why. It is a shame that the idea of being conservative has been wholly swallowed in contemporary discourse by big-C political Conservatism. But I think there's a way out for a notion of conservatism that we should all want to hang onto, no matter our political persuasion. That is the idea that I've been developing throughout this book—that adopting a conservative disposition toward our interactions with the world is about trying to achieve the ability to manage change on terms that recognize our agency and allow for the realization of participatory goods.[38] Concerns about appropriation offer a window on how this process can go wrong, and highlight the often heightened consequences for those who are already marginalized, but the lessons we find there are relevant to all of us. In what ways do you actually exercise control over the things that you care about preserving? What do you take for granted because it just happens that something you value has been well maintained up to this point? We could all benefit from being more involved in the conservation of things we care about, not only because of the benefits of participation but because the more we shirk the work of preserving things, the more we give up control. And once we give up control, we may be forced to confront change on terms that we do not want to accept. That isn't just a problem for some specific groups or communities—it's a problem for all of us.

7 | FROM OURSELVES TO FUTURE GENERATIONS

FOR WHOM SHOULD WE SAVE THINGS?

Remember our friends at Engineered Labs? They were the ones whose viral TikTok of the apparent smashing of Indus Valley terra cotta elicited outrage for its casual and disrespectful attitude toward the care of South Asian material heritage (even though the pottery in the video turned out to be fake). Another reason that the video caused such a furor is that Engineered Labs sells a product that they call the "Heritage Personal Museum," a collection of fragments from thirty-three historical artifacts, cast in acrylic: from "Pompeii Ash" to "Dinosaur egg shells." Considering this alongside the video, it looked like the company was deliberately breaking items in order to divide them up and sell them.[1] A representative of the company insists this is not the case, and I have no way of verifying whether the items in the "Heritage Personal Museum" are any more authentic than the Indus Valley terra cotta replica. But the product does nicely capture an important aspect of common thinking about heritage that in turn plays a key role in conservation—the idea that there are important parts of a common human heritage that anyone might be interested in, and therefore anyone might justifiably possess. To say that our heritage is held in common is to say, on this view, that we all have a claim to it. The "Heritage Personal Museum" is just the "Encyclopedic Universal Museum" (the museum that aspires to collect representative samples of artifacts from all times and places) writ small.

The previous chapter might suggest an implicit view on which all of the myriad and wonderful things in the world should always and only be conserved on behalf of particular discrete groups of people that stand in special identity-based relations to them. (After all, we have surveyed the robust connections that often link participation in the conservation of objects, practices, and places, with core dimensions of individual and group identity.) But I don't think that at all. On the contrary, many of the world's rich cultural, artistic, and environmental goods have a universal character to them—they are things that anyone could have good reason to enjoy, care about, or marvel at. The idea of a universal cultural heritage has historically been abused, employed as a veneer for the imposition of Eurocentric value systems and the exclusion of marginalized perspectives. For this reason, the idea of a universal cultural heritage has been criticized for many years and from many quarters, and much of that criticism is well-taken, as we'll soon see. But these problems should be laid at the feet of those who have misused and exploited the idea of universal value—properly understood, they are not objections to the concept itself. It turns out that the idea of universal value needs saving, too.

To see this, it is crucial to distinguish questions about who should *control* or *possess* conservation candidates from who has reason to *value* them. Throughout the previous chapters of this book, we've seen the ways that participation in conservation projects on terms that allow for the achievement of participatory goods for relevant parties has a key part to play in our conservation thinking. But in the principles undergirding the notion of the "universal museum" and other policies and institutions connected to the universal value paradigm, we often see the conflicting assumption that a commitment to the universal value of particular goods should entail that they are controlled and owned by certain actors—namely, the well-resourced, often Western, museums that currently happen to possess many of them. At the very least, according to this perspective, it's viewed as generally *permissible* for such institutions to own and control these objects. We should reject that blanket assumption. We can be committed to both the repatriation and redistribution of cultural goods in the possession of colonial powers *and* the idea that these goods should, subject to various culturally specific exceptions, be available for the enjoyment of everyone.[2] Rather than thinking that retention by encyclopedic museums is the best way to facilitate appreciation of these goods, we should turn our attention to novel policies and norms that would encourage the sharing of the fantastically diverse objects of wonder in the world on more equitable terms.

It's helpful to begin by contrasting two different ways of thinking about the "universal" in universal value—one that is singular, and one that is multifaceted. According to the singular account, to say that something (say, Yellowstone National Park or the Taj Mahal, to pick two UNESCO World Heritage sites) has universal value is to say that everyone shares the same reason to value it. So, for example, we might say that Yellowstone exhibits a kind of natural splendor or Taj Mahal a kind of architectural magnificence that anyone might rightly appreciate. From this perspective, it's as if the universally valuable thing in question has a single lock, and each of us possesses a copy of the matching key should we wish to use it. According to the multifaceted account, in contrast, to say that something has universal value is to say that anyone would rightly value *that* thing, but this might be for all manner of different reasons. Maybe you appreciate the raw power of Old Faithful but I gravitate toward the arrestingly still palette of Grand Prismatic Spring; maybe you marvel at the scale of Taj Mahal and I am moved by its story. From this perspective, it's as if the universally valuable thing has an untold number of locks, and people have all manner of different keys that might well fit a lock and allow them access to an aspect of its value.[3]

Proponents of the singular approach might criticize the multifaceters by questioning whether their conception of universal value results in an understanding of these goods that is universal in the right way. Rhetoric and policy surrounding the idea of World Heritage, for example, are often couched in the language of *shared* value and *common* heritage. For example, the 1982 UNESCO World Heritage listing document states: "Their [the sites'] value cannot be confined to one nation or to one people, but is there to be shared by every man, woman and child of the globe."[4] We might then think that in order to fit the bill, the way that we think about universal value needs to account for a shared understanding of these goods, and a recognition that they are valuable precisely because they are part of a heritage we hold in common—*that* is the single, shared reason that we have for thinking of their value as universal. The multifaceted account doesn't capture this essential idea of a shared heritage.

But is it not enough that a particular object, tradition, or place is the shared focus of our interest, even if we all bring a radical diversity of approaches to understanding its value? I worry that the proponent of the singular view is overly influenced by the idea of a common heritage, a kind of value that is supposed to achieve universality by stroke of our shared humanity.[5] This is a component of the approach that we saw espoused in Appiah's cosmopolitan critique of cultural patrimony back in Chapter 5, the idea of a kind of value that we come to appreciate specifically *despite difference.* I agree that there is something ennobling about this conception of universal value, relying as it does on an understanding of goods that, despite their radical plurality, have a value that transcends *our* differences and reaches for its source in some attribute that unites us all, including those who have died and those yet to be born. But this criterion is also narrowing, limiting our resources for approaching such values through an emphasis on features that we share. It does not preclude that we may also find our own idiosyncratic meanings in these goods, but it does demand that what makes them qualify as universal is the common ground we share for valuing them.[6] I resist the idea that universal value is the sole province of Ralph Waldo Emerson's transparent eye-ball, a good that we can access only by transcending all that makes us *us.*

I am happy to grant that this kind of singular universal value may be a distinctive kind of good (I'm not *anti*-transcendence), but we should not require that our general understanding of universal value be forced into its mold. There is also something distinctive and marvelous about things that offer a radical plurality of approaches to appreciating them. I find art and literature to offer a better model for universal value here than UNESCO World Heritage sites, both with respect to the diversity of universally valuable things and the diversity of ways to appreciate them. Some examples will help us see why.

On the first score, we might reasonably ask: How do we reconcile the incredible variety of valuable things in the world with the idea of universal value? Isn't there supposed to be something *special* about universal value (as encyclopedic museums and UNESCO World Heritage sites might suggest) that would be eroded if we claimed that there was so much of it out there? In a recent discussion of universalism in aesthetics, the philosopher Alex King argues that these two ideas actually go hand in hand. There are in fact so many worthwhile things to appreciate in the world that no one person in our finite lives could possibly come close to scratching the surface. Diversity of interest in the abundance of the world's value helps us all get just a little bit closer to experiencing more of it, because we can be each other's guides to the artworks and other objects of aesthetic interest that we haven't yet experienced, or heard about, or don't yet have enough context or knowledge to fully appreciate. But the fact that there are so many amazing things in the world shouldn't lead us to call into question that they are each universally valuable, the kind of thing that anyone might appreciate—it should, rather, reinforce our impression that the world is really that awesome.[7]

On the second score, one of the wonders of shared appreciation for art is precisely that we *don't* necessarily share the same reasons for valuing it, that it can speak to us in different ways, not *no matter* who we are, in the literal sense, but *because of* who we are, whoever that may be. The philosopher Nick Riggle has recently articulated a view of aesthetic conversation that is oriented around *community* rather than *convergence*.[8] According to the predominant convergence view, when we discuss (or argue about!) what we find aesthetically worthwhile, we are aiming to come to an agreement: I think the novel is moving, whereas you find it maudlin, and in talking it through, we're shooting for the elimination of tension between our views. In contrast, according to the community view, we are aiming for a "mutual valuing of individuality."[9] We don't need to aim at reconciling our divergent aesthetic judgments, but can rather try to appreciate each other's particular perspective on the same object of attention.

Although this account is specific to *aesthetic* conversation, we can extract an important lesson for how we think about the general idea of universal value. Rather than thinking that the core of universality is convergence on a shared source of value, the kind of consideration that pushes us toward the transparent eye-ball perspective of the singular account, we can think of it as rooted in a global and timeless community: a rowdy, crowded house in which we each bring a different view to the table and we aim to recognize what we each find valuable in the same thing. There is so much more to learn from the different perspectives we all bring to a common object of attention than can be found in the preemptive demand for a single reason that we can all share in common. Moreover, the multifaceted approach honors our common humanity by recognizing one of our most important attributes as a species—our capacity for an especially radical form of autonomous agency. In this, I am more in line with comments that Appiah makes elsewhere: "Cosmopolitanism imagines a world in which people and novels and music and films and philosophies travel between places where they are understood differently, because people are different and welcome to their difference. Cosmopolitanism can work because there can be common conversations about these shared ideas and objects."[10] This ability to direct our lives in dialogue with history, tradition, and community grants us unique perspectives to share with one another. That variety should be at the core of our thinking, even when we're focused on values that unite us.

The multifaceted approach does entail that many things hold the promise of universal value, but I find this a welcome and accurate consequence. I don't think it leads to the proliferation problem, discussed with reference to uniqueness and irreplaceability in Chapter 3, where we are forced to accept that *everything* has universal value—the value of mementos and keepsakes will, appropriately, tend to be partial and limited, like G. A. Cohen's eraser. It's not inconceivable that everyone might find something to value about a dingy old eraser—but it also isn't likely. And it's not the kind of object that anyone is likely to *present* as a candidate for universal appreciation. In the end, there remains something aspirational about the idea of universal value.[11] It's not a category where membership is secured by an empirical test—we don't determine it by surveying representative samples of the population about whether they think something is worth caring about, hoping for unanimity. Rather, when we think that an object, practice, or place is special in a way that promises to offer value for anyone, we can treat it accordingly, conferring on it this aspirational status as a candidate for universal value.[12] And the open secret is that it ultimately doesn't matter if something's value is *really* universal in this sense, actually something that literally anyone might find valuable. This is part of what makes the criterion of "Outstanding Universal Value" employed by UNESCO for World Heritage designation somewhat ridiculous—as if some committee is going to ferret out which proposals *actually* meet the bar.[13] Clearly even the most ardent defender of World Heritage or the universal museum would not be shocked or dismayed if someone came along and said "Meh . . ." to their candidate for universal value, as if its mere rejection by *someone* would dethrone it from its universal status. To treat something as universally valuable is to honor it, and each other, by recognizing it as something we believe is *worth* anyone's time, whether or not everyone actually has time for it.[14]

With this multifaceted model of universal value in mind, we're in a better position to address the legitimate concerns that have been leveled against the ways that the universal paradigm has been employed. For example, consider this quote from Caribbean-British cultural theorist Stuart Hall: "who is the Heritage for? In the British case the answer is clear. It is intended for those who 'belong'—a society which is imagined as, in broad terms, culturally homogeneous and unified."[15] Hall's criticism of how "British Heritage" is conceived is that unity is achieved only at the cost of excluding those who don't already "fit in," what political theorists call a "stipulative unanimity": it's like taking a vote, kicking out all the nay votes, and claiming there was unanimous support for the yeas.

This kind of concern has a physical manifestation when it comes to universal value rhetoric in nature conservation. Nature conservation is typically presented as being for the benefit of all, whether we're thinking about it in an environmentalist light (we all depend on the health of the planet!) or an appreciative one (natural beauty should be available for all to experience!). But when it comes to the kind of fortress conservation often found in the establishment of national parks, for example, the practices involved in their creation give the lie to the universal value rhetoric. Parks such as Yellowstone and Yosemite in the United States or Kruger in South Africa were created through the forced displacement of Indigenous peoples from their ancestral homes.[16] It can be especially hard to stomach the language of "preservation on behalf of all!" in the face of such practices, especially when the facts on the ground demonstrate that access to these places remains highly stratified on the basis of race and class.[17]

In both its symbolic and literal manifestations, then, the thrust of criticisms surrounding appeals to universal value is that they are harmfully exclusionary. The critic's argument goes like this: appeals to universal value are just a front for a particular set of partial values (e.g., Eurocentric ones) or a scam that an action benefits everyone when it only benefits the elite.[18] But those are consequences we should reject, and therefore we should reject appeals to universal value in turn. That is certainly *a* solution to the problem, but a preferable approach, I submit, is to demand that appeals to universal value actually aspire to universality, to demand that they do aim to capture goods that anyone might care about, or that they do redound to the benefit of all. If we break the link between universal value and exclusion, then the aspiration of universality can remain standing.

In a landmark report on heritage conservation from the Getty Foundation, the authors write: "Universality—the assumption that some heritage is meaningful to all of mankind—is one of the basic assumptions and matters of faith underlying conservation practice and one assumption that emphasizes the positive role of heritage in promoting unity and understanding."[19] Here, as in our discussion of universal value earlier, we can see unity as an aspiration, a goal of coming together *across* difference rather than the rejection of difference. But how do we achieve that kind of goal if our conception of universality is bound by the singular paradigm? The singular approach ends up laying the groundwork for the imposition of a value system embraced by those in power. If the demand is that we all must value our common heritage for the same reason, then *no wonder* the reason typically on offer reflects the values of dominant cultures! The multifaceted approach, in contrast, *invites* a multiplicity of perspectives—by not insisting on a shared value, it can avoid the game of competition and ranking. In doing so, it opens up space for us to discover the universal character of objects, practices, and places that don't fit the Eurocentric "universal" value mold at all.

Earlier, we saw that critics of universal value tend to reject the idea altogether on the basis of the purported link between universal value claims and Eurocentric values. A similar argumentative structure is found in worries about a claimed connection between the idea of universal value and a universal right to possession or control. If such a right is entailed by appeals to universal value, and we want to reject such a claim (for any number of reasons, including those discussed in the last chapter), then it would seem that we need to reject the idea of universal value, too. That's again a logically valid argument, but it's not the only route to solving the problem. Another approach is to sever the supposed entailment from universal value to universal rights—if these two ideas don't need to go hand in hand, then we can reject universal rights to possession and control while salvaging some conception of universal value.

The multifaceted approach has offered us tools for understanding the kind of *attitudes* that universal value invites, but what of rights and actions? If universal value doesn't entail universal rights to access, possession, or control, then what does it imply about what we should *do*? An alternative model can be found in the notion of a universal *responsibility*.[20] Some of our universal responsibilities have a scope that extends far beyond conservation candidates that we might purport to have universal value. For example, a general responsibility to not wantonly destroy what other people value does not require that what they value have anything approaching a universal character to it. In fact, you may be the only person in the world who cares about a small rock cairn in your local park, but even that gives the rest of us good reason not to maliciously or carelessly kick it over.[21] Now, to be sure, the reason here is quite weak: there are all manner of considerations that might tell against the preservation of your rock cairn. Maybe it's smack in the middle of a walking path, or between first and second base where the kids play softball. This kind of thin responsibility to not destroy what other people care about isn't going to get us very far when it comes to hard cases, but by that stroke, it also shouldn't be very controversial.

For conservation candidates whose value has a universal character, by contrast, more powerful reasons to save them might emerge. Protected areas such as national parks offer an instructive model—if we think that anyone has reason to value natural wonders, then we might posit not only that we have reason not to carelessly destroy these places but also that we have reason not to destroy them for profit, or for building schools, or any number of other considerations, noble or not, that might compete with them for space. Put another way, we might think that we have reason not just to avoid destroying, but to actively protect and preserve natural wonders.[22]

We need not assume that the strength of our reason to preserve things scales with the number of people who value them according to some set function in order to recognize that things with the potential to matter to lots of people will tend to call out for more active and intentional preservation projects. This is partially for pragmatic reasons (we're liable to anger a lot of people if we let the stuff everyone cares about fall into disrepair), but it's plausibly for moral reasons as well. When we neglect the objects, practices, and places that give shape and meaning to people's lives, we disrespect them. This is worst when we willfully suppress, erase, or destroy what people care about, but it also cuts when we cannot lift a finger (or a penny) to help save them.

—

But who is the "we" in the previous paragraphs? The reason not to willfully destroy what others cherish is a moral consideration that applies to all of us as individuals—it's part and parcel of our reason not to be bullies. To help others with their preservation projects, on the other hand, would be a nice thing to do, but it's not a moral failure to spend your personal time in other ways. You may be deeply committed to cleaning up the local park or volunteering at the arts council, but I don't err if I prefer to spend my Sundays doing the crossword with my wife and playing video games with my kid. So, who is the "we" that fails in neglecting preservation work?

Sometimes having a responsibility to support a good does not mean we have an obligation to do certain tasks in support of it ourselves. Indeed, sometimes it would be impossible or counter-productive for us to try to pursue them on our own. Rather, they are goods that we pursue collectively. Public education is a crucial good, but the importance of pursuing it doesn't entail that we all have a responsibility to be public K–12 teachers (though that is an excellent thing to do). But we do have a responsibility to collectively provide for public education, and there are a range of ways that an individual can help fulfill that obligation. We might pay our taxes, vote for a measure, campaign for it, volunteer some time, make a donation. These are familiar ways that we can meet our responsibilities to support public goods.

The defining features of public goods are typically that they be "non-excludable" (available to all) and "non-rivalrous" (my enjoying the good doesn't preclude or diminish your enjoying it as well). Clean air is a classic example—if the air is clean in a place, everyone there has access to it, and no one's breathing of the air limits the availability of it to others.[23] Now, not all goods that we designate as public will fully meet these conditions, and public parks are a good example—they're not strictly non-rivalrous goods, because when the park gets extremely crowded, that can in fact preclude or diminish its enjoyment by others (a lesson that denizens of many cities felt pointedly during the COVID-19 pandemic).[24] But goods that we think should be for everyone certainly have a public character to them, and, like public parks, it makes sense that public institutions have an important role to play in their care and maintenance. They can do this through a variety of means—through laws and ordinances that offer special protections, through local, state, and federal granting agencies, through public education, and so forth.

However, there is already inadequate public funding for preservation projects, and the account of universal value I've advocated for here suggests that there are even more conservation candidates with a universal character than we might initially assume. This presents a serious practical problem, but not, I think, any evaluative error. The mistake, rather, is to assume that there is an overly limited set of conservation candidates with universal value, and to think that we can discharge our responsibilities to support conservation by protecting these things alone.

It's important to recognize that terms such as "everyone" and "universal" can be relativized to more local communities. When we traffic in these concepts, we often reach for the grand scale of national parks and World Heritage sites, but it would be a mistake to think that the same patterns of reasoning don't apply to the park in a small town. The town park is also for anyone/everyone, and so, too, aspires to a kind of universality that is captured by its public character; in a way, a town can be a whole world. This doesn't mean that the town park isn't also preserved for the sake of visitors, and isn't also a place that everyone beyond the scope of the town might appreciate, too, but we can distinguish the potential reach of its value from its target audience.

Although it may at first seem oxymoronic, when we come to recognize the universal in the local, we can map out a more promising route to pursuing a broader set of conservation projects. Given the radical diversity of things worth saving, if we each devote some attention to the things we care about, we will collectively engage in a kind of grassroots conservation of many things.

There is still an important role for public funding in such projects, and we can look to models such as the U.S. National Endowment for the Humanities' division of Preservation and Access or the National Trust for Historic Preservation for examples of how federal or national programs can aid local preservation projects (though to be sure, the NEH remains sorely underfunded). But what's crucial to see is that local conservation projects need not be special-interest projects.[25] If we devote too much of our conservation attention to the anointed candidates favored by a singular universal value ideal, the World Heritage sites and national parks of the world, we can lose site of the universal appeal that can be found in the local, the humble, and the everyday. The multifaceted understanding of universal value helps us to see this potential. Consider these sentiments in the preamble to the 1954 Hague Convention for the Protection of Cultural Property in the Event of Armed Conflict:

> Being convinced that damage to cultural property belonging to any people whatsoever means damage to the cultural heritage of all mankind, since each people makes its contribution to the culture of the world;
>
> Considering that the preservation of the cultural heritage is of great importance for all peoples of the world and that it is important that this heritage should receive international protection.[26]

While language geared toward an international audience tends to take national groups as its unit of analysis, the reasoning employed in the Hague Convention is just as applicable to culture that has not been recognized on the national or international scale, whether because it is from small, local communities, or because it has been overlooked or ignored. It is not only the cultural objects or practices sanctioned by a nation that contribute to the heritage of all humankind but also the mural in a neglected post-industrial city, the regional cuisine known only to locals, the grasslands that have yet to ignite the imagination of anyone with the power to designate national parks.[27] The better we can provide the support for people to show how what they care about has a universal character—is worth anyone's attention—the better we protect our collective resources for understanding ourselves and each other.[28]

We will surely fail at preserving much of what is valuable in the world. This is simply the consequence of recognizing how much is worth saving and the limitations we inevitably face in protecting it. But the better that we, collectively, provide amenable terms for people to manage losses and changes to the things that they value, the better we allow for the preservation of meaning and the preservation of identity, even when the world around us is in flux.

So far, we've focused on whom conservation is for along an axis that we might construe as *spatial*—of all the people in this wide world, for which of them do we conserve things? As we've seen, we think some conservation candidates approach a universal status that suggests that the answer is *everyone*. But in addition to this spatial axis, there is also a *temporal* axis that raises an analogous question, and which has long been a concern of conservationists, sometimes superseding a focus on the interests of the current generation. For example, the Victorian writer John Ruskin, a famously conservative voice in architectural conservation, wrote: "*We have no right whatever to touch them*. [The buildings] are not ours. They belong partly to those who built them, and partly to all the generations of mankind who are to follow us. The dead have still their right in them."[29] William Murtagh, the first Keeper of the National Register of Historic Places, put things less dogmatically: "It has been said that, at its best, preservation engages the past in a conversation with the present over a mutual concern for the future."[30] Even in more personal contexts, the idea that we are saving things for the future is familiar. We don't just conserve family heirlooms for our own sakes, but for that of future family members as well. We "pass down" traditions as well as objects, so that aspects of our culture are not completely erased by the forces of change.[31]

When references to preserving things for the future surface in policy documents, such as the World Heritage Convention, those appeals are often vague.[32] We are told that we should save things "for future generations," but who are these future generations? How far in the future are we talking about? Who is to say that these future generations will care at all about what we care about now? If we think that saving things for the future is the primary justification for preservation work, we can quickly run into a problem. One heritage professional interviewed about conserving things for the future lamented that if at some point everyone might lose interest in what they're laboring to conserve, then what's the point?[33] This concern is premised on the logic that preservation is always and only for the sake of the future; but that line of thinking is doomed from the start, because there's nowhere for the buck to stop. Yes, eventually the sun will explode and any remaining humans (optimistically) will be disintegrated back into stardust—surely *that* doesn't undermine the point of preservation. But if the certainty of there *eventually* being no one left to care about what we labor to conserve now doesn't make such pursuits frivolous, then why think the possibility of such an occurrence one hundred years out would do so either?

The mistake is to think that the future is always just beyond the next hill and out of sight, and not also what happens in the next year, or week, or day. When we talk about preserving things for future generations, we can certainly have in mind people one hundred years out and beyond, but we shouldn't think of this time horizon to the exclusion of more proximate ones.[34] If we successfully conserve something that people cherish for the next one hundred years, why should it undermine the point of that work if the day after one hundred years everyone stops caring?[35] We might lament that change in values, or object to it depending on the terms on which the change transpires, but it doesn't render the conservation work worthless.

It's also important to remember that the conservation work we undertake now lays the groundwork for what future generations even have the option of valuing. It's possible that future generations won't care about Indus Valley terra cotta or Kansas City tacos or performing the haka, but if we let these things be lost or destroyed or forgotten, then we preclude the possibility of future generations being *able* to value them. Preservation work doesn't need to be hitched to the future certainty of people caring about what we care about—but it does create the possibility of our sharing those values with people yet to come. Throughout this book, we've seen the ways that preservation can shape people's sense of self, their conception of who they are. Preservation for future generations can in turn shape the space of who future people might be by maintaining particular materials from which an identity can be constructed. If our identity is always built up from the things that matter to us, then what is available for us to care about in the first place will inevitably constrain or expand the possibilities for understanding ourselves.

Crucially, this does not require that future generations value the same things that we do, nor that they be required to force their sense of identity into the molds created by our own contemporary categories and their combinations and permutations. But what we do by conserving things that we value is offer resources that future generations can take up as materials in forming their own practices and habits—materials to play with, adjust, alter, and tailor.

One last common trope from fiction is instructive here. There's a familiar plot move in fantasy and sci-fi where it is discovered that the ancestors of the current generation achieved glorious feats of art and science that were lost in one kind of cataclysmic event or the other.[36] We empathize with the characters as they slowly uncover the marvels (and sometimes the horrors) of knowledge and skills that they never had the opportunity to inherit. This setup is so powerful because it speaks to the deep sense of loss engendered by a lack of access to these forgotten resources, not only pragmatically but artistically, culturally, morally—the stuff out of which we craft our sense of self. And just as this is a task that we pursue in part together with others synchronically, so, too, is it something we do in dialogue with others diachronically.

The philosopher Sam Scheffler argues that traditions, like personal routines, are a way of carving out a home in time analogous to a home in space. But whereas personal routines are just for you, traditions also offer us the opportunity to belong to something larger than ourselves.[37] Recall that in Chapter 5, we considered how sharp disruptions in the character of a place can produce strangeness and noted that habit is a tool for establishing and maintaining familiarity.[38] Routines and traditions "domesticate" time by making it familiar; we walk these paths, say these words, cook these foods, and in so doing, we take something as terrifying as an undifferentiated and uncontrolled expanse of time and make it welcoming and approachable.[39] We divest it of strangeness by carving out a place where we can feel at home.

These reflections on familiarity and strangeness can cast the idea of saving things for the future in a different light. Preservation can provide a temporal continuity that retains familiarity analogous to the spatial familiarity of retaining the character of a place. We recognize this song, this smell, this landmark, this keepsake, and even though many other aspects of the world may change, we remain anchored to the past and are not simply swept away. It's no surprise that people who move or are displaced to radically different environments tend to cluster together into cultural enclaves where the sights, sounds, and scents of their homelands are reproduced and provide a sense of stability, the materials that ground a continued grip on identity, both individual and shared. What serves in space can serve equally well in time.

As I write these words, I glance at the wall: hanging from a nail, there is a thick leather strap studded with brass cowbells that descend in size from top to bottom. The bells are from the farm in Pennsylvania where my grandmother grew up and where my father visited as a child, where generations of that branch of my family tree had lived. Next to the bells, there is a framed newspaper clipping of the 1958 advertisement for the farm's public sale, including machinery, household goods, and a herd of twenty-seven Holsteins. The cows are listed in detail along with their names: Dell, Nelma, Pearl, Babe, Rose, Polly, Shirley Mae. Lunch available. Terms—Cash.

With the exception of a brief summer job in high school, farming is alien to me, and I've never worked with cows. I don't value these vestiges of that life in the way that my forebears did. But my grandmother saved these things, and then my father did—so I save them, too. They are part of the furniture of our home; familiar, but simultaneously a portal to a time and place that I don't fully understand. Maybe they will ultimately be meaningless to my daughter, who never knew my grandmother or my father. But they aren't yet. Her second-grade class did family history projects in November, and she chose to write about the cowbells. We save the bells and in doing so, in some small part, we can spy the branch of the family tree that our stem shoots off from. We glimpse a different branch when we light Yahrzeit candles on Yom Kippur, another still when we leaf through the Hatala family Polish cookbook. When we confront intergenerational time, we always grapple with a tension between the familiar and the strange, what we glean from stories and objects, and what remains difficult to grasp or even fathom. We choose to save these things—our daughter may choose otherwise. That will be her choice. But by saving what we do, what we find meaningful, beautiful, or sobering, we give her that choice to make. And so it is with everything that we preserve for the future. Generations down the line may not care about the same things we do or in the same way, may not conserve what we have labored to conserve. But along with the things that we save, we preserve the choice about whether to continue the project of conservation. Conserving things helps secure the terms that empower future generations to decide for themselves what they will save—who they will be.

NOTES

CHAPTER 1

1. For some helpful discussion of these questions, see Scarbrough (2022).
2. Lefebvre (2019).
3. Scheffler (2007, 106).
4. This is a version of what philosophers call *the paradox of hedonism*.
5. For further discussion of how the past specifically is made important through the perception that it is threatened, see Lowenthal (1985).
6. I am inspired by a quote that will be familiar to philosophers. Wilfrid Sellars wrote: "The aim of philosophy, abstractly formulated, is to understand how things in the broadest possible sense of the term hang together in the broadest possible sense of the term" (1962, 37).
7. For a helpful overview of the history and theory of conservation as a set of professional fields, see Muñoz Viñas (2011). For a careful look at how conservation work proceeds in the domains of architecture, painting, and literary/historical editing (and the relations among them), see Eggert (2009).
8. For an example of philosophical work that takes up some of these questions, see Carrier (1985). Irvin (2022) helpfully explores how the changing nature of some contemporary artworks (and the rules that accompany them) might catalyze rethinking of how to answer these questions (see Ch. 3, in particular). For a wide-ranging discussion of some of these questions from an art historical and conservation perspective, see Scott (2016).
9. This case is also discussed in Irvin (2022, 59). For philosophical discussion of what the goals of art restoration ought to be, see de Clercq (2013).

10. For discussion of Facadism, see Schumacher (2010); Fisher (2020).
11. Bicknell (2014, 438)
12. Korsmeyer (2008, 123).
13. Walter (2021, 10–11).
14. Norton (1986).
15. In this I follow Hölling (2017).
16. As quoted in Merryman (1986).
17. While this commitment is not shared among all approaches to ethical theory, it is found in a wide range of literatures, reaching back at least to Aristotle in the Western philosophical tradition.
18. Muñoz Viñas (2020, 7).
19. Cantalamessa (2020).
20. Muñoz Viñas elsewhere acknowledges a role for intersubjectivity, but he seems to regard it as mere agreement among a group of people, rather than a goal of rational discourse. See Muñoz Viñas (2011, 150–53).

CHAPTER 2

1. Steinbeck (1939, 88).
2. This extension of the case was introduced by Thomas Hobbes in *De Corpore* (Molesworth 1994, 136–37).
3. Steinbeck (1939, 88–89).
4. Smith (2006, 44).
5. Cohen (2011, 221). The link that Cohen invokes between caring about the past and the very nature of our humanity is a common touchstone among the late greats of 20th-century moral philosophy. As Joseph Raz puts it: "To deny our past is to be false to ourselves. This is justification enough for our dependence on our past" (2001, 34). Or more starkly, consider these words from Stuart Hampshire: "Persons who conspicuously enjoy and excel in reasoning, but who have no interest in any kind of story-telling or in recalling and recording their past, tend to be considered monsters of rationality, and be called inhuman" (1989, 44).
6. For further discussion, see Matthes (2013).
7. Korsmeyer (2012, 372). See also Harold (2020).

8. Kagan (1998).
9. Korsmeyer et al. (2014, 432).
10. Mapes (2017).
11. Korsmeyer et al. (2014, 432).
12. Riegl (1982).
13. L. Turner (2016).
14. For discussion, see Matthes (2017c).
15. Benjamin (2008).
16. Korsmeyer (2016, 224). For further discussion, see Matthes (2018a).
17. Holtorf (2005, 118) as cited in S. Jones (2010, 183).
18. The rest of this paragraph follows the discussion in Dutton (1979).
19. Irvin (2007). For further discussion of the van Meegeren case, see Scott (2016, Ch. 4).
20. Muñoz Viñas discusses the expectation-dependent nature of authenticity in Muñoz Viñas (2020, 24–27). He identifies this as a "problem" for authenticity discourse, highlighting that it "makes the notion more volatile than it may seem." In contrast, I argue here that understanding that authenticity is expectation-dependent is helpful for sorting through competing claims about authenticity.
21. Han (2017).
22. For discussion, see Riggs (2021).
23. Strohl (2019, 159).
24. Muñoz Viñas (2020, 28).
25. Elliot (1982); Katz (1992).
26. Vogel (2015).
27. Dowie (2011).
28. Strohl (2019, 161–62). Strohl also employs the example of Viet-Cajun barbecue.
29. Brammer (2019).
30. Bialystok (2011). Charles Lindholm's description of authenticity appears to capture both dimensions in turn: "Authentic objects, persons, and collectives are original, real, and pure; they are what they purport to be, their roots are known and verified, their essence and appearance are one" (2008, 2).

31. Cf. S. Jones (2010, 187).
32. Heldke and Thomsen (2014).
33. As S. Jones puts it, "the process of negotiating the authenticity of material things can also be a means of establishing the authenticity of the self" (2010, 189).
34. For discussion, see Lindholm (2008), especially Part II.
35. Ríos-Hernández (2022).
36. P. Taylor (2016, Ch. 5). The idea that our heritage is defined in part by its use for present and future purposes is a familiar idea in the contemporary heritage studies literature. See, for example, Smith (2006, 44); Harrison (2013, 14).
37. Light (2000). For a discussion of repairing objects as a point of departure for repairing relationships, see Spellman (2002).
38. Matthes (2017c).
39. Bluestone (2011, 14–15).
40. Bluestone (2011, 18).
41. Brady (2002, 78).
42. For discussion of this phenomenon in the context of urban homogenization, see Nguyen (2022).
43. The archaeologists Siân Jones and Thomas Yarrow write: "the fundamental paradox of conservation . . . [is] how to keep things in some essential way the same, even as they and the world transforms [*sic*]" (2022, 2).
44. For further discussion, see Matthes (2020).
45. Killian (1998).
46. For further discussion, see Matthes (2018b).
47. Osborne (2017).
48. Yuko (2021).
49. As the philosopher David Velleman puts it: "particular cares and concerns can be definitive of a person's identity or essential to the self" (2002, 112).
50. Whyte (2018, 131).
51. Whyte (2018, 131).
52. This idea is simpatico with Korsgaard's account of practical identity and normativity in the case of individuals, a topic we'll return to in the following chapter.

53. Whyte (2017). On cultural heritage as a constituent element of flourishing, see Harding (1999).

CHAPTER 3

1. Stegner (1969).
2. For a comprehensive history of wilderness protection in the United States, see J. Turner (2013).
3. Moore (1903).
4. Gruen (2002). Compare also with S. Jones (2017, 26): "It might be preferable to conceive of social value as a *process of valuing* heritage places rather than a fixed value category that can be defined and measured."
5. McShane (2007, 53).
6. Matthes (forthcoming).
7. C. M. Korsgaard (1983); Langton (2007).
8. Cf. McShane (2007, 58).
9. Saito (2022, 101).
10. I originally introduce the proliferation problem in Matthes (2013). See also Siegel (2022, 144): "the collapse of a limited canon of art opens up not only a vast range of possible art to appreciate but a limitless amount of art to protect or mourn."
11. Cohen (2011); Martin (1979); Kolodny (2003); O'Neill, Holland, and Light (2008).
12. Cf. Raz (2001, 28, n15).
13. Kagan (1998, 283); Sibley (2001).
14. Danto (1965, 167).
15. Thanks to the participants in Elizabeth Marlowe's virtual heritage studies book club for drawing my attention to this case, not to mention for prompting me to read a range of fascinating books that I draw on throughout these pages.
16. Matthes (2013).
17. The U.S. Flag Code may be found online at https://uscode.house.gov/view.xhtml?path=/prelim@title4/chapter1&edition=prelim.
18. Karlstrom (2015, 38-39).
19. Ferguson, Anyon, and Ladd (2000).

20. Cf. Anderson (1995, 26); Matthes (2019b). For further discussion, see DeSilvey (2017).
21. Giannini (2023).
22. For discussion, see James (2011; 2013).
23. Onibada (2022).
24. Lenzen (2022).
25. We might think of this in relation to McGowan's account of how speech can enact permissibility-facts that oppress, though in these cases it's a video rather than speech per se. See McGowan (2009).
26. Allen (2022).
27. Cf. Giombini (2022), who claims that we never owe respect to the object but only to the communities for whom it is meaningful.
28. Cf. Li (2021).
29. Holtorf (2015) mentions that the destruction of the 20th century has led to a huge number of archaeological discoveries. For discussion of the relationship between repair and destruction, see Spellman (2002, Ch. 7).
30. Raz (2001). This is in conflict with Giombini's claim that the kind of respect relevant to conservation is only respect in treatment (2022, 102).
31. For example, see "Council Conclusions of 21 May 2014 on Cultural Heritage as a Strategic Resource for a Sustainable Europe" (2014).
32. "Text of the Convention for the Safeguarding of the Intangible Cultural Heritage" (2003).
33. Cross, n.d.
34. Giombini (2022, 105) describes the essential role that a place might play in helping to constitute a people's identity as being "irreplaceable."
35. E.g., Smith (2006).
36. Harding (1999); James (2019).
37. C. Korsgaard (1996).
38. My aim is to put Korsgaard's notion of practical identity into dialogue with the moral psychology of "felt needs" discussed in the next chapter, not to follow her in offering an overall account of the nature of moral normativity.
39. Cf. Williams (1981).

40. Hassoun and Wong (2015); Etieyibo (2017); e.g., Sluss and Ashforth (2007).
41. Cf. "Life is this self-making, autopoietic, but fundamentally xenophobic activity of producing, preserving, conserving, and insisting on the integrity of oneself" (Noë 2022, 123).
42. S. Goodman (2023). See also Nowak (2019).
43. For a more stringent view, Kyle Whyte mentions a quote from Frances Van Zile to the effect that without wild rice and water, the Anishinaabe people would cease to be Anishinaabe at all (2018, 132).
44. For a detailed history of the South Natick Dam, see Diamant (2021).
45. Brown (2022).
46. https://www.savenatickdam.org/.
47. Fox, Magilligan, and Sneddon (2016). Thanks to Jillian Wilson-Martin for bringing this article to my attention.
48. As quoted in Gerstenblith (2016, 342).
49. Norton (1986).
50. Compare Hassoun and Wong (2015, 117): "relationalism tells us to preserve nature as part of what makes us who we are or could be."

CHAPTER 4

1. Schulz (2022, 5).
2. Korsmeyer (2022, 88).
3. A. Q. Thompson (2022).
4. Riggs (2021, 30).
5. Holland and Rawles (1994, 46); O'Neill, Holland, and Light (2008).
6. O'Neill, Holland, and Light (2008, 157); Arntzen (2008); Scoville (2013).
7. Arntzen (2003; 2008). For further discussion, see Saito (2007, 146).
8. A related emphasis on process rather than product is discussed in relation to some Buddhist temples in Laos by Karlstrom (2015, 38).
9. Ferguson, Anyon, and Ladd (2000).
10. O'Neill, Holland, and Light (2008, 2–3); J. Thompson (2000).
11. Whyte (2018, 138).

12. Walter thinks it is an advantage of narrative accounts that they are subject to this contestation: that's a reason *to* include many different perspectives (Lamarque and Walter 2019). For discussion, see Fisher (2020, 95). For further discussion of competing narratives in conservation, see Scarbrough (2020).
13. See, for example, Walter (2021, 80); Hoesch (2022, 8); Brady (2002, 89).
14. Brady (2002, 89). Brady refers to this kind of integrity as "diachronic" in contrast with "synchronic," which roughly maps onto Arntzen's distinction between dynamic and static preservation (Arntzen 2008). The references to "wholeness" and "continuity" draw directly from Holland and O'Neill (1996).
15. Brady (2002, 89).
16. Yarrow (2018, 339).
17. Yarrow (2018, 339).
18. Cf. Eggert (2009, 107).
19. Yarrow (2018, 338). On Yarrow's gloss: "In the former, the destruction is through the loss of originality through material degradation through the passage of time; in the latter it is through conservator's own 'artificial' intervention."
20. For helpful discussion of the role of participation in shaping character in the case of offshore wind farms, see Saito (2017, 106–8).
21. Nomikos (2018, 456). For further discussion of familiarity and strangeness in environmental aesthetics, see Brady (2011); Paola and Ciccarelli (2022).
22. Haapala (2005, 46).
23. Hourdequin and Havlick (2011, 78).
24. McShane (2012).
25. Toolis (2017, 185–87).
26. I am grateful to my former student Refilwe Kotane for a class project that brought this issue to my attention.
27. Toolis (2017).
28. For further discussion of the value of self-determination in a political context, but with broader implications, see Stilz (2015).

29. Whyte (2018).
30. Arnaquq-Baril (2016).
31. Scheffler (2007, 108).
32. For further discussion, see Hoesch (2022); Jeffers (2013, 506).
33. By comparison, Hoesch outlines two criteria for acceptable change in terms of authenticity and continuity (8), though this conception of authenticity has much in common with the exercise of autonomous agency. For helpful discussion of the relationship between authenticity and autonomy, see Oshana (2007). The continuity criterion concerns the *rate* of change (i.e., not too fast).
34. Cf. Lear (2008) .
35. Denis, Hummel, and Pontille (2022). For a related point on the burdens placed on us by physical objects that recalls their status as witnesses (discussed in Chapter 2), consider this passage from Hilary Mantel's *The Mirror and the Light*: "Our possessions outlast us, surviving shocks that we cannot; we have to live up to them, as they will be our witnesses when we are gone" (2020, 64). For more on the aesthetic dimensions of the imposition of objects upon us, see Saito (2022, 54ff).
36. Wonderly (2016, 228; 2021, 157).
37. Wonderly (2021, 155).
38. Wonderly (2021, 171).
39. Wonderly (2016, 234).
40. Compare with Shoemaker (2003, 114): "when I care for something, I find part of myself tied tightly to it, such that when something bad happens to it, something bad (emotionally) happens to me. So when I am moved to preserve something I care about, I am acting to preserve my investments. In other words, I am acting on behalf of myself, or at least the part of myself that is tied to the cared-for object. But this is just to say that I am, in such circumstances, determined and authorized to act *by my self*."
41. Wonderly (2021, 163).
42. Watene (2016, 293).
43. Jacobs (1995).

44. Figueroa and Waitt (2010, 144).
45. Climbing at Uluru was ultimately officially prohibited.
46. Ypi (2017).
47. Onuzo (2016).
48. For a simpatico discussion about how letting a destroyed site ruinate can allow it to serve as a reminder of the profound losses occasioned by war, see Scarbrough (2020).
49. E. L. Thompson (2022).
50. Compare with Irvin on some works of contemporary art: "In a context where the preservation of objects in a preferred condition has long been the central aspiration of conservation, the choice to allow deterioration has particularly strong expressive import" (2022a, 75).
51. "This Pig Wants to Party: Maurice Sendak's Latest" (2011).
52. Singer (2023).
53. Quiroz (2022).
54. For further discussion of such protests in the context of civil disobedience, see Lai and Lim (2023). For discussion of how some contemporary artworks engage in this kind of transgression, see Irvin (2022a), which I will discuss further in Chapter 5.
55. For further philosophical discussion of the Bamiyan Buddhas, see Janowski (2011); Bülow and Thomas (2020).
56. For example, al Manzali (2016).
57. Simons (2016).
58. McKenzie (2017).
59. For discussion, see Frowe and Matravers (forthcoming).
60. Cf. hooks (1995, xv).
61. Compare with Whyte (2017).
62. S. Goodman (2023).
63. Jeffers (2015).
64. See also Taylor (2016, Ch. 5).
65. Gross (2023).
66. Boxill (1976).
67. For example, see Shelby (2002).

CHAPTER 5

1. The discussion of 'alalā in this and the subsequent paragraph follows van Dooren (2016).
2. Compare with Noë (2022)'s discussion of David Brower's comments concerning conserving the California condor. Brower thought this required preserving the condor's environment, and Noe extends this point to the arts, as I discuss later in the chapter. But note the constraints on that approach suggested by van Dooren's reflections on 'alalā preservation—we should be wary of insisting that the changes birds experience in an altered environment render them essentially compromised from a conservation perspective.
3. Eaton and Gaskell (2022, 59); Lamarque and Walter (2019); Henderson (2022, 105).
4. Compare: "According to the people quoted above preservation means cultural preservation: the active maintenance of continuity with indigenous values and beliefs that are part of a community's identity" (Clavir 2002, 73).
5. Yarrow (2019, 12).
6. Yarrow (2019, 13).
7. Ingold (2010). For discussion, see Holtorf (2015).
8. Domínguez Rubio (2016, 66).
9. Domínguez Rubio (2016, 61–62).
10. Tuan (1977, 6).
11. For relevant discussion, see Dawdy (2016).
12. Domínguez Rubio (2016, 64). The philosopher Alva Noë makes a similar point: "Real conservation," he writes, "must therefore be a kind of dedication or commitment to the whole system of relationships, which is the artwork" (2022, 126).
13. The use of this term is not meant to prejudge whether species conservation should take place in situ or ex situ.
14. Merryman (1994, 64). Merryman contrasts an "object-oriented" cultural property policy with a "nation-oriented" one. See also Warren (1989) on collecting cultural property. In addition to articulating a conflict

resolution model for cultural property disputes, she ultimately points toward preservation of non-renewable resources such as endangered species as a productive framework.

15. This prioritization is echoed in Clavir's description of dominant views among curators: "Many conservators, while agreeing on the importance of the information, believe in the immediate value of preserving objects as an end in itself" (2002, 28).
16. Clavir (2002).
17. For rich discussion of whether paintings originally intended for non-museum contexts (such as altarpieces) can be preserved in museums, see Carrier (1985; 2001).
18. Cheam-Shapiro (2023).
19. Noë (2022, 126).
20. Riggle (2010).
21. Riggle (2010, 248).
22. Digital alteration makes it look like Katsu is tagging a Picasso; you can find it on YouTube. Banksy's film *Exit through the Giftshop* features some of his guerilla additions to the collections of major museums. Tom Green performed a similar stunt on *The Tom Green Show*, adding his painting "Tiger Zebra" to the walls of The National Gallery. Again, you can YouTube it.
23. Laver (1936, 121).
24. "there is always the question of whether the way they are conserved and presented might undermine their very authenticity by cutting them (and us) off from the unique networks of relationships they embody" (S. Jones 2010, 200).
25. See Raz (2001, 163) on preservation as a preliminary to engagement with value.
26. University College London (2023).
27. "No War Can Destroy These Minecraft-Made Heritage Sites" (2019).
28. Elliot (1982).
29. L. Turner (2016).
30. J. Jones (2016). For further discussion, see Matthes (2017c; 2018).

31. Korsmeyer (2022). For further discussion on this point in the context of aesthetic appreciation, see Korsmeyer (2016; 2019); Matthes (2018a).
32. Matthes (2017a).
33. Yarrow (2019, 7).
34. S. Jones et al. (2018).
35. S. Jones et al. (2018, 344).
36. S. Jones et al. (2018, 346). Compare: "people use authenticity to work out genuine or truthful relationships between objects, people and places, and this process is heightened by the forms of dislocation and displacement that characterize the modern world" (S. Jones 2010, 198).
37. J. Taylor and Gibson (2017); Rafii (2023, 242).
38. Rafii (2023, 236).
39. Rafii (2023, 239).
40. Hourdequin and Havlick (2011, 74).
41. Whyte (2016, 127). Whyte, collective food relations.
42. For recent discussion of participatory approaches to conservation in natural resource management, see Matarrita-Cascante, Sene-Harper, and Ruyle (2019). For discussion and examples of participatory heritage conservation, see Peters et al. (2020). For relevant discussion of the related notion of participatory justice in contexts from solar-radiation management to food, see Hourdequin (2019); Loo (2014). For a recent examination of PAR (Participatory Action Research), see Lenette (2022). For discussion of participation in adapting to loss, see Venture et al. (2021).
43. Killmister (2011, 242).
44. Compare with the discussion of Fred and authentic motorcycle restoration in Spellman (2002, 13–15).
45. For discussion of Monique Wonderly's work on attachment and felt need, see Chapter 5.
46. Sandoval, Rudhru, and Ser (2016, 512); Sandoval (2016, 512).
47. Mingon and Sutton (2021, 4360).
48. Yarrow (2019, 7).
49. Stewart (2022).
50. For discussion of the aesthetic dimensions of caring for things yourself, see Saito (2022, 141).

51. Appiah (2006, 135).
52. Cf. Matthes (2015). Appiah's discussion of "rooted Cosmopolitanism" in *The Ethics of Identity* includes some discussion more in line with this idea. See Appiah (2007, Ch. 6).
53. Brady (2016, 100).
54. Cochrane (2009, 72). See also Polite (2019).
55. Raz et al. (2005, 27–28).

CHAPTER 6

1. See Redman (2016).
2. For example, see Kaufman (2009); Lepore (2021).
3. Though I've tried. See Matthes (2016; 2019a). As I argue in those papers, I don't think cultural appropriation is a unified and unique kind of wrong, but is a descriptive label that can capture a heterogeneous set of wrongs related to marginalization. Some of those wrongs are discussed in the remainder of this chapter. For other work on the general topic of cultural appropriation, see, for example, J. O. Young (2005; 2008); J. O. Young and Brunk (2012); Coombe (1993); Rogers (2006); Nguyen and Strohl (2019).
4. For an overview, see Harjo (1996).
5. On social group as neither aggregates nor associations, see I. M. Young (2011).
6. For a similar point in literary representation, see Shim (2021).
7. Ferguson, Anyon, and Ladd (2000). For an interesting case of Zuni artists shaping representation through map-making, see the A:shiwi A:wan Museum and Heritage Center.
8. For an overview, see Grasswick (2018).
9. Valdman (2009); Liberto (2014).
10. See Sène (2022); Dowie (2011); Murdock (2021).
11. See also O'Neill (2002).
12. Horse Capture (2018).
13. Norby (2023; emphasis added).
14. Meissner (2018, 272).
15. Meissner (2018, 274).

16. Meissner (2020, 357).
17. Irvin (2022b).
18. Elgin (1993, 17).
19. Irvin (2022b, 66).
20. N. Goodman (1968, 52).
21. Irvin (2022b, 73).
22. Irvin (2022b, 72).
23. Irvin (2022a, 65).
24. This is a more precise version of the idea that cultural appropriation can be wrongful when it manifests oppressive relations. See Matthes (2019a).
25. This argument is simpatico with literature about participatory justice in a range of fields. See, for example, Hourdequin (2019); Loo (2014); Matarrita-Cascante, Sene-Harper, and Ruyle (2019); Lenette (2022).
26. For relevant discussion, see Nguyen and Strohl (2019).
27. J. O. Young (2005, 136); Ziff and Rao (1997, 3).
28. For an argument about the importance of legal property frameworks to the case of intangible cultural goods, see Carpenter, Katyal, and Riley (2009).
29. For further discussion of such cases, see Sax (2001).
30. This point is indebted to Dorceta Taylor's Douglas Lecture at Wellesley College, 2021.
31. Gilbert (2023).
32. Icon (2023).
33. For further discussion, see E. L. Thompson (2022).
34. Osborne (2017); Lim (2020).
35. "Declaration of the Importance and Value of Universal Museums" (2004).
36. Kaufman (2009, 32).
37. Wylie (2005). For further critical discussion in environmental context, see Palmer (1992).
38. In this I depart from Cohen's use of the term "conservative disposition" in Cohen (2011). His conception is about favoring bearers of value over value maximization; mine is about the terms on which such choices are confronted.

CHAPTER 7

1. Onibada (2022).
2. For further discussion, see Matthes (2017b).
3. For further discussion, see Matthes (2015; forthcoming). In those texts, I use the terms "monistic" and "pluralistic" rather than "singular" and "multifaceted."
4. As quoted in Omland (2006, 247).
5. For discussion, see Fabre (forthcoming).
6. Compare Pantazatos: "Each local participant at a World Heritage site is able to make her/his own understanding, meaning and interpretation of the site. The ascription of outstanding universal value does not constrain participants from making their own associations with World Heritage sites" (2017, 381).
7. King (forthcoming, 10–11).
8. Riggle (2021).
9. Riggle (2021, 642).
10. Appiah (2007, 258).
11. This differs from the model of aspirational agency that Callard describes in Callard (2018). For in conferring the status of universal value, we ourselves already value the thing in question: our aspiration is on behalf of others.
12. There is some affinity here with aspects of the Institutional Theory of art. See Dickie (2008).
13. For a detailed study of how UNESCO World Heritage decisions are made, see Meskell (2018).
14. Compare with Scheffler (2010b).
15. Hall (1999, 6).
16. Dowie (2011); Meskell (2011).
17. Cf. Murdock (2021, 239).
18. Compare Matherne (2022, 455) for discussion of a Kantian model of aesthetic humility.
19. Avrami, Mason, and De La Torre (2000, 69).
20. Cf. "A common World Heritage can be interpreted in terms of a shared global moral obligation to protect the cultural heritage of all peoples of the world" (Omland 2006, 245).

21. Raz (2001).
22. It's worth emphasizing, though, that their preservation need not be premised on a complete ban on human use. Once we give up on the sharp distinction between nature and culture, there's no reason to assume that all human uses of protected areas are inconsistent with their conservation.
23. Hume (forthcoming).
24. Miller (2004); Lindsay (2012).
25. Cf. Hume (forthcoming): "In short, activities whose proximate benefits are private, in some salient respect, may still ultimately contribute towards commonly valuable, non- excludable goods."
26. Gerstenblith (2016, 347).
27. For an argument that all old artifacts share a weak presumption in favor of their preservation on the basis of their history as part of our social world, see Sher (2023).
28. For a parallel argument in the context of public funding for the arts, see Hume (forthcoming).
29. Ruskin, as quoted in Eggert, 26. From "Lamp of Memory," 358.
30. Murtagh (1988, 168).
31. For a comprehensive discussion of tradition, see Shils (2006).
32. Högberg et al. (2017, 640).
33. Högberg et al. (2017, 643).
34. Cf. Henderson (2022).
35. Compare with Scheffler (2013). I think the point applies somewhat in the imagined scenario where we *assume* that others will stop caring after 100 years, but it applies strongly where we are only entertaining that possibility.
36. See, for example, The Wheel of Time or The Stormlight Archive.
37. Scheffler (2010a).
38. Haapala (2005); Nomikos (2018).
39. Scheffler (2010a).

REFERENCES

Allen, Amy. 2022. "Feminist Perspectives on Power." In *The Stanford Encyclopedia of Philosophy*, edited by Edward N. Zalta and Uri Nodelman. Metaphysics Research Lab, Stanford University. https://plato.stanford.edu/archives/fall2022/entries/feminist-power/.

Anderson, Elizabeth. 1995. *Value in Ethics and Economics*. Cambridge, MA: Harvard University Press.

Appiah, Kwame Anthony. 2006. *Cosmopolitanism*, 115–35. New York: W. W. Norton.

Appiah, Kwame Anthony. 2007. *The Ethics of Identity*. Princeton, NJ: Princeton University Press.

Arnaquq-Baril, Alethea, dir. 2016. *Angry Inuk*. National Film Board of Canada.

Arntzen, Sven. 2003. "Whose City, Whose Environment? Self-Determination, Ethics and the Urban Environment." In *Place and Location III*, edited by Virve Sarapik and Kadri Tüür, 55–74. Tallinn: Proceedings of the Estonian Academy of Arts.

Arntzen, Sven. 2008. "The Complex Cultural Landscape: Humans and the Land, Preservation and Change." In *Humans in the Land: The Ethics and Aesthetics of the Cultural Landscape*, edited by Emily Brady and Sven Antzen, 39–74. Oslo: Oslo Academic Press, Unipub Norway.

Avrami, Eric C., Randall Mason, and Marta De La Torre. 2000. "Values and Heritage Conservation: Research Report." Los Angeles, CA: Getty Conservation Institute. https://www.getty.edu/conservation/publications_resources/pdf_publications/values_heritage_research_report.html.

Benjamin, Walter. 2008. *The Work of Art in the Age of Its Technological Reproducibility, and Other Writings on Media*. Edited by Michael W. Jennings, Brigid Doherty, and Thomas Y. Levin. Cambridge, MA: Belknap Press of Harvard University.

Bialystok, Lauren. 2011. "Refuting Polonius: Sincerity, Authenticity, and 'Shtick.'" *Philosophical Papers* 40 (2): 207–31. https://doi.org/10.1080/05568641.2011.591816.

Bicknell, Jeanette. 2014. "Architectural Ghosts." *Journal of Aesthetics and Art Criticism* 72 (4): 435–40.

Bluestone, Daniel. 2011. *Buildings, Landscapes, and Memory: Case Studies in Historic Preservation*. New York: W.W. Norton.

Boxill, Bernard R. 1976. "Self-Respect and Protest." *Philosophy and Public Affairs* 6 (1): 58–69.

Brady, Emily. 2002. "Aesthetic Character and Aesthetic Integrity in Environmental Conservation." *Environmental Ethics* 24: 75–91.

Brady, Emily. 2011. "The Ugly Truth: Negative Aesthetics and Environment." *Royal Institute of Philosophy Supplement* 69: 83–99. https://doi.org/10.1017/s1358246111000221.

Brady, Michael S. 2016. "Group Emotion and Group Understanding." In *The Epistemic Life of Groups: Essays in the Epistemology of Collectives*, edited by Michael Brady and Miranda Fricker, 95–110. Oxford: Oxford University Press.

Brammer, John Paul. 2019. "I'm from a Mexican Family. Stop Expecting Me to Eat 'Authentic' Food." *Washington Post*, May 20, 2019. https://www.washingtonpost.com/outlook/2019/05/15/im-mexican-american-stop-expecting-me-eat-authentic-food/.

Brown, Bob. 2022. "Select Board Votes 4-to-1 to Remove South Natick Dam & Waterfall." *Natick Report*. November 10, 2022. https://www.natickreport.com/2022/11/natick-select-board-votes-4-to-1-to-remove-south-natick-dam-waterfall/.

Bülow, William, and Joshua Lewis Thomas. 2020. "On the Ethics of Reconstructing Destroyed Cultural Heritage Monuments." *Journal of the American Philosophical Association* 6 (4): 483–501. https://doi.org/10.1017/apa.2020.11.

Callard, Agnes. 2018. *Aspiration: The Agency of Becoming*. Oxford: Oxford University Press.

Cantalamessa, Elizabeth. 2020. "Debating Bon Jovi's Cheesiness Will Enrich Your Conceptual Life | Psyche Ideas." *Psyche*, October 20, 2020. https://psyche.co/ideas/debating-bon-jovis-cheesiness-will-enrich-your-conceptual-life.

Carpenter, Kristen A., Sonia K. Katyal, and Angela R. Riley. 2009. "In Defense of Property." *Yale Law Journal* 118 (6): 1022–125.

Carrier, David. 1985. "Art and Its Preservation." *Journal of Aesthetics and Art Criticism* 43 (3): 291–300.

Carrier, David. 2001. "Art Museums, Old Paintings, and Our Knowledge of the Past." *History and Theory* 40: 170–89.

Cheam-Shapiro, Sophiline. 2023. "Met Museum Kicked Me Out for Praying to My Ancestral Gods." *Hyperallergic*. March 21, 2023. http://hyperallergic.com/809442/met-museum-kicked-me-out-for-praying-to-my-ancestral-gods/.

Clavir, Miriam. 2002. *Preserving What Is Valued*. Vancouver: UBC Press.

Cochrane, Tom. 2009. "Joint Attention to Music." *British Journal of Aesthetics* 49 (1): 59–73. https://doi.org/10.1093/aesthj/ayn059.

Cohen, G. A. 2011. "Rescuing Conservatism: A Defense of Existing Value." In *Reasons and Recognition: Essays on the Philosophy of T. M. Scanlon,* edited by R. Jay Wallace, Rahul Kumar, and Samuel Freeman, 203–31. New York: Oxford University Press.

Coombe, Rosemary J. 1993. "The Properties of Culture and the Politics of Possessing Identity: Native Claims in the Cultural Appropriation Controversy." *Canadian Journal of Law and Jurisprudence* 6 (2): 249–85.

"Council Conclusions of 21 May 2014 on Cultural Heritage as a Strategic Resource for a Sustainable Europe." 2014. *Official Journal of the European Union* 183: 36–38. https://eur-lex.europa.eu/legal-content/EN/TXT/PDF/?uri=CELEX:52014XG0614(08)&from=FR.

Cross, Anthony. n.d. "Social Aesthetic Goods and Aesthetic Alienation." *Philosophers' Imprint*. https://doi.org/10.3998/phimp.3475

Danto, Arthur C. 1965. *Analytical Philosophy of History*. Cambridge: Cambridge University Press.

Dawdy, Shannon Lee. 2016. *Patina: A Profane Archaeology*. Chicago: University of Chicago Press.

DeSilvey, Caitlin. 2017. *Curated Decay: Heritage Beyond Saving*. Minneapolis: University of Minnesota Press.

"DOCUMENT: Declaration on the Importance and Value of Universal Museums." 2006. In *Museum Frictions: Public Cultures/Global Transformations* edited by Ivan Karp, Corinne A. Kratz, Lynn Szwaja and Tomas Ybarra-Frausto, 247–49. New York: Duke University Press. https://doi.org/10.1515/9780822388296-015

De Clercq, Rafael. 2013. "The Metaphysics of Art Restoration." *British Journal of Aesthetics* 53 (3): 261–75. https://doi.org/10.1093/aesthj/ayt013.

Denis, Jérôme, Cornelia Hummel, and David Pontille. 2022. "Getting Attached to a Classic Mustang: Use, Maintenance and the Burden of Authenticity." *Journal of Material Culture* 27 (3): 259–79. https://doi.org/10.1177/13591835211068940.

Diamant, Charlotte. 2021. "The Environmental History of the South Natick Dam." https://repository.wellesley.edu/object/ir1341

Dickie, George. 2008. "What Is Art? An Institutional Analysis." In *Aesthetics: A Comprehensive Anthology*, edited by Steven M. Cahn and Aaron Meskin, 426–37. Malden, MA: Blackwell.

Domínguez Rubio, Fernando. 2016. "On the Discrepancy between Objects and Things: An Ecological Approach." *Journal of Material Culture* 21 (1): 58–86.

van Dooren, Thom. 2016. "Authentic Crows: Identity, Captivity and Emergent Forms of Life." *Theory, Culture and Society* 33 (2): 29–52. https://doi.org/10.1177/0263276415571941.

Dowie, Mark. 2011. *Conservation Refugees*. Cambridge, MA: MIT Press.

Dutton, Denis. 1979. "Artistic Crimes: The Problem of Forgery in the Arts." *British Journal of Aesthetics* 19 (4): 302–14.

Eaton, A. W., and Ivan Gaskell. 2022. "Introduction: Active Matter—Some Initial Philosophical Considerations." In *Conserving Active Matter*, edited by Peter N. Miller and Soon Kai Poh, 51–64. New York: Bard Graduate Center.

Eggert, Paul. 2009. *Securing the Past*. Cambridge: Cambridge University Press.

Elgin, Catherine Z. 1993. "Understanding: Art and Science." *Synthese* 95 (1): 13–28. https://doi.org/10.1007/bf01064665.

Elliot, Robert. 1982. "Faking Nature." *Inquiry* 25: 81–93.

Etieyibo, Edwin. 2017. "Ubuntu, Cosmopolitanism, and Distribution of Natural Resources." *Philosophical Papers* 46 (1): 139–62. https://doi.org/10.1080/05568641.2017.1295616.

Fabre, Cécile. forthcoming. *To Snatch Something from Death: Value, Justice, and Humankind's Common Cultural Heritage*. Edited by M. Matheson. The Tanner Lectures on Human Values. Salt Lake City: University of Utah.

Ferguson, T. J., Roger Anyon, and Edmund J. Ladd. 2000. "Repatriation at the Pueblo of Zuni: Diverse Solutions to Complex Problems." In *Repatriation Reader: Who Owns American Indian Remains?*, edited by Devon A. Mihesuah, 239–65. Lincoln: University of Nebraska Press.

Figueroa, Robert Melchior, and Gordon Waitt. 2010. "Climb: Restorative Justice, Environmental Heritage, and the Moral Terrains of Uluru-Kata Tjuta National Park." *Environmental Philosophy* 7 (2): 135–63.

Fisher, Saul. 2020. "Lifespans of Built Structures, Narrativity, and Conservation: A Critical Note." *Estetika: The European Journal of Aesthetics* LVII/XIII, no. 1: 93–103. https://doi.org/10.33134/eeja.31.

Fox, Coleen A., Francis J. Magilligan, and Christopher S. Sneddon. 2016. "'You Kill the Dam, You Are Killing a Part of Me': Dam Removal and the Environmental Politics of River Restoration." *Geoforum* 70 (March): 93–104. https://doi.org/10.1016/j.geoforum.2016.02.013.

Frowe, Helen, and Derek Matravers. forthcoming. *Stones and Lives*. Oxford: Oxford University Press.

Gerstenblith, Patty. 2016. "The Destruction of Cultural Heritage: A Crime against Property or a Crime against People?, 15 J. Marshall Rev. Intell. Prop. L. 336 (2016)." *UIC Review of Intellectual Property Law* 15 (3): 336–93. https://repository.law.uic.edu/ripl/vol15/iss3/1.

Giannini, Melissa. 2023. "Searching for Meg White." *ELLE*. June 1, 2023. https://www.elle.com/culture/music/a43846386/meg-white-interview-2023/.

Gilbert, Margaret. 2023. *Life in Groups: How We Think, Feel, and Act Together*. Oxford: Oxford University Press.

Giombini, Lisa. 2022. "Respect in Conservation Ethics: A Philosophical Inquiry." *Studies in Conservation* 67 (1–2): 100–108. https://doi.org/10.1080/00393630.2021.1940797.

Goodman, Nelson. 1968. *Languages of Art: An Approach to a Theory of Symbols*. Indianapolis: Bobbs-Merrill.

Goodman, Sylvia. 2023. "Native American Languages Are Disappearing: Colleges Could Help Preserve Them." *Chronicle of Higher Education*. January 23, 2023. https://www.chronicle.com/article/native-american-languages-are-disappearing-colleges-could-help-preserve-them.

Grasswick, Heidi. 2018. "Feminist Social Epistemology." In *The Stanford Encyclopedia of Philosophy*, edited by Edward N. Zalta. Metaphysics Research Lab, Stanford University. https://plato.stanford.edu/archives/fall2018/entries/feminist-social-epistemology/.

Gross, Terry. 2023. "Drag Queen (and Ordained Minister) Bella DuBalle Won't Be Silenced by New Tenn. Law." *NPR*, March 16, 2023, sec. Law. https://www.npr.org/2023/03/16/1163815547/tennessee-drag-law-queen-bella-duballe.

Gruen, Lori. 2002. "Refocusing Environmental Ethics: From Intrinsic Value to Endorsable Valuations." *Philosophy and Geography* 5 (2): 153–64. https://doi.org/10.1080/10903770220152380.

Haapala, Arto. 2005. "On the Aesthetics of the Everyday: Familiarity, Strangeness, and the Meaning of Place." In *The Aesthetics of Everyday Life*, edited by Andrew Light and Jonathan M. Smith, 39–55. New York: Columbia University Press.

Hall, Stuart. 1999. "Whose Heritage? Un-Settling 'The Heritage,' Re-Imagining the Post-Nation" *Third Text* 13 (49): 3–13. https://doi.org/10.1080/09528829908576818.

Hampshire, Stuart. 1989. *Innocence and Experience*. Cambridge, MA: Harvard University Press.

Han, Byung-Chul. 2017. *Shanzhai*. Cambridge, MA: MIT Press.

Harding, Sarah. 1999. "Value, Obligation and Cultural Heritage." *Arizona State Law Journal* 291: 291–354.

Harjo, Suzan Shown. 1996. "Introduction." *In Mending the Circle: A Native American Repatriation Guide*, edited by Barbara Meister, 3–7. New York: American Indian Ritual Object Repatriation Foundation.

Harold, James. 2020. *Dangerous Art: On Moral Criticisms of Artwork. Thinking Art*. Oxford; New York: Oxford University Press.

Harrison, Rodney. 2013. *Heritage: Critical Approaches*. New York: Routledge.

Hassoun, Nicole J., and David B. Wong. 2015. "Conserving Nature; Preserving Identity." *Journal of Chinese Philosophy* 42 (1–2): 176–96. https://doi.org/10.1111/1540-6253.12189.

Heldke, Lisa, and Jens Thomsen. 2014. "Two Concepts of Authenticity." *Social Philosophy Today* 30: 79–94.

Henderson, Jane. 2022. "Conservators Delivering Change." *Studies in Conservation* 67 (sup1): 105–11. https://doi.org/10.1080/00393630.2022.2066320.

Hoesch, Matthias. 2022. "The Right to Preserve Culture." *Journal of Moral Philosophy* 1 (aop): 1–26. https://doi.org/10.1163/17455243-20223607.

Högberg, Anders, Cornelius Holtorf, Sarah May, and Gustav Wollentz. 2017. "No Future in Archaeological Heritage Management?" *World Archaeology* 49 (5): 639–47. https://doi.org/10.1080/00438243.2017.1406398.

Holland, Alan, and John O'Neill. 1996. "The Integrity of Nature over Time Some Problems." In *The Thingmount Working Paper Series on the Philosophy of Conservation*, 1–18. Lancaster: Lancaster University, Department of Philosophy.

Holland, Alan, and Kate Rawles. 1994. "The Ethics of Conservation." In *The Thingmount Working Paper Series on The Philosophy of Conservation*, 1–56. Lancaster: Lancaster University, Department of Philosophy.

Hölling, Hanna. 2017. "The Technique of Conservation: On Realms of Theory and Cultures of Practice." *Journal of the Institute of Conservation* 40 (2): 87–96. https://doi.org/10.1080/19455224.2017.1322114.

Holtorf, Cornelius. 2005. *From Stonehenge to Las Vegas: Archaeology as Popular Culture*. Lanham, MD: Rowman Altamira.

Holtorf, Cornelius. 2015. "Averting Loss Aversion in Cultural Heritage." *International Journal of Heritage Studies* 21 (4): 405–21.

hooks, bell. 1995. *Art on My Mind*. New York: The New Press.

Horse Capture, Joe. 2018. "Horse Capture: 'Native People Have a Story to Tell—Their Own.'" *ICT News*. September 13, 2018. https://ictnews.org/archive/horse-capture-native-people-have-a-story-to-tell-their-own.

Hourdequin, Marion. 2019. "Geoengineering Justice: The Role of Recognition." *Science, Technology, and Human Values* 44 (3): 448–77. https://doi.org/10.1177/0162243918802893.

Hourdequin, Marion, and David G. Havlick. 2011. "Ecological Restoration in Context: Ethics and the Naturalization of Former Military Lands." *Ethics, Policy and Environment* 14 (1): 69–89.

Hume, Jack Alexander. 2023. "Neutrality, Cultural Literacy, and Arts Funding." *Ergo: An Open Access Journal of Philosophy* 10 (55): 1588–1617.

Icon. 2023. "Leading with Passion: Coordinating a Conservation Team of Volunteers with Ages from 16–90." May 25, 2023. https://www.icon.org.uk/resource/leading-a-conservation-team-of-volunteers-with-ages-from-16-90.html.

Ingold, Tim. 2010. "No More Ancient; No More Human: The Future Past of Archaeology and Anthropology." In *Archaeology and Anthropology*, edited by D. Garrow and T. Yarrow, 160–70. Oxford: Oxbow.

Irvin, Sherri. 2007. "Forgery and the Corruption of Aesthetic Understanding." *Canadian Journal of Philosophy* 37 (2): 283–304.

Irvin, Sherri. 2022a. *Immaterial: Rules in Contemporary Art*. Oxford: Oxford University Press.

Irvin, Sherri. 2022b. "The Expressive Import of Degradation and Decay in Contemporary Art." In *Conserving Active Matter*, edited by Peter N. Miller and Soon Kai Poh, 65–79. New York: Bard Graduate Center.

Jacobs, Michael. 1995. "Sustainability and Community: Environment, Economic Rationalism and the Sense of Place." *Australian Planner* 32 (2): 109–15. https://doi.org/10.1080/07293682.1995.9657669.

James, Simon P. 2011. "For the Sake of a Stone? Inanimate Things and the Demands of Morality." *Inquiry: An Interdisciplinary Journal of Philosophy* 54 (4): 384–97. https://doi.org/10.1080/0020174x.2011.592343.

James, Simon P. 2013. "Why Old Things Matter." *Journal of Moral Philosophy* 12 (3): 313–29.

James, Simon P. 2019. "Natural Meanings and Cultural Values." *Environmental Ethics* 41 (1): 3–16. https://doi.org/10.5840/enviroethics20194112.

Janowski, James. 2011. "Bringing Back Bamiyan's Buddhas." *Journal of Applied Philosophy* 28 (1): 44–64. https://doi.org/10.1111/j.1468-5930.2010.00512.x.

Jeffers, Chike. 2013. "Appiah's Cosmopolitanism." *Southern Journal of Philosophy* 51 (4): 488–510.

Jeffers, Chike. 2015. "The Ethics and Politics of Cultural Preservation." *Journal of Value Inquiry* 49 (1–2): 205–20.

Jones, Jonathan. 2016. "Palmyra Must Not Be Fixed: History Would Never Forgive Us." *Guardian*.

Jones, Siân. 2010. "Negotiating Authentic Objects and Authentic Selves." *Journal of Material Culture* 15 (2): 181–203.

Jones, Siân. 2017. "Wrestling with the Social Value of Heritage: Problems, Dilemmas and Opportunities." *Journal of Community Archaeology and Heritage* 4 (1): 21–37. https://doi.org/10.1080/20518196.2016.1193996.

Jones, Siân, Jeffrey Stuart, Mhairi Maxwell, Alex Hale, and Cara Jones. 2018. "3D Heritage Visualisation and the Negotiation of Authenticity: The ACCORD Project." *International Journal of Heritage Studies* 24 (4), 333–53.

Jones, Siân, and Thomas Yarrow. 2022. *The Object of Conservation: An Ethnography of Heritage Practice*. London: Routledge.

Kagan, Shelly. 1998. "Rethinking Intrinsic Value." *Journal of Ethics* 2 (4): 277–97.

Karlstrom, Anna. 2015. "Authenticity." In *Heritage Keywords*, edited by Kathryn Lafrenz Samuels and Trinidad Rico, 29–46. Denver: University Press of Colorado.

Katz, Eric. 1992. "The Big Lie: Human Restoration of Nature." *Research in Philosophy and Technology* 12: 231–41.

Kaufman, Ned. 2009. *Race, Place, and Story*. New York: Taylor and Francis.

Killian, Ted. 1998. "Public and Private, Power and Space." In *The Production of Public Space*, edited by Andrew Light and Jonathan M. Smith, 115–34. Lanham, MD: Rowman & Littlefield.

Killmister, Suzy. 2011. "Group-Differentiated Rights and the Problem of Membership." *Social Theory and Practice* 37 (2): 227–55.

King, Alex. forthcoming. "Universalism and the Problem of Aesthetic Diversity." *Journal of the American Philosophical Association*, 1–20. https://doi.org/10.1017/apa.2022.53.

Kolodny, Niko. 2003. "Love as Valuing a Relationship." *Philosophical Review* 112 (2): 135–89.

Korsgaard, Christine M. 1983. "Two Distinctions in Goodness." *Philosophical Review* 92 (2): 169–95.

Korsgaard, Christine. 1996. *The Sources of Normativity*. New York: Cambridge University Press.

Korsmeyer, Carolyn. 2008. "Aesthetic Deception: On Encounters with the Past." *Journal of Aesthetics and Art Criticism* 66 (2): 117–27.

Korsmeyer, Carolyn. 2012. "Touch and the Experience of the Genuine." *British Journal of Aesthetics* 52 (4): 365–77.

Korsmeyer, Carolyn. 2016. "Real Old Things." *British Journal of Aesthetics* 56 (3): 219–31.

Korsmeyer, Carolyn. 2019. *Things: In Touch with the Past*. Oxford: Oxford University Press.

Korsmeyer, Carolyn. 2022. "The Look of Age: Appearance and Reality." In *Conserving Active Matter*, edited by Peter N. Miller and Soon Kai Poh, 80–99. New York: Bard Graduate Center.

Korsmeyer, Carolyn, Jennifer Judkins, Jeanette Bicknell, and Elizabeth Scarbrough. 2014. “Symposium: The Aesthetics of Ruin and Absence.” *Journal of Aesthetics and Art Criticism* 72 (4): 429–49.

Lai, Ten-Herng, and Chong-Ming Lim. 2023. “Protest and Cultural Artefacts.” *Public Ethics*. April 5, 2023. https://www.publicethics.org/post/protest-and-cultural-artefacts.

Lamarque, Peter, and Nigel Walter. 2019. “The Application of Narrative to the Conservation of Historic Buildings.” *Estetika: The European Journal of Aesthetics* LVI/XII, no. 1: 5–27.

Langton, Rae. 2007. “Objective and Unconditioned Value.” *Philosophical Review* 116 (2): 157–85. https://doi.org/10.1215/00318108-2006-034.

Laver, James. 1936. “Museums or Mausoleums.” *History* 21 (82): 120–30.

Lear, Jonathan. 2008. *Radical Hope: Ethics in the Face of Cultural Devastation*. Cambridge, MA: Harvard University Press.

Lefebvre, Sam. 2019. “‘This Is Reparations:’ S.F. School Board Votes to Paint Over Controversial High School Mural.” *KQED*. June 25, 2019. https://www.kqed.org/arts/13860237/this-is-reparations-s-f-school-board-votes-to-paint-over-controversial-high-school-mural.

Lenette, Caroline. 2022. *Participatory Action Research: Ethics and Decolonization. Research to the Point*. Oxford: Oxford University Press.

Lenzen, Cecilia. 2022. “‘They Never Respected Us’: TikToker Criticized after Intentionally Breaking 3,000-Year-Old Pottery Artifact.” *Daily Dot*. May 12, 2022. https://www.dailydot.com/irl/tiktoker-breaks-pottery-artifact/.

Lepore, Jill. 2021. “When Black History Is Unearthed, Who Gets to Speak for the Dead?” *New Yorker*, September 27, 2021. https://www.newyorker.com/magazine/2021/10/04/when-black-history-is-unearthed-who-gets-to-speak-for-the-dead.

Li, Zhen. 2021. “Immorality and Transgressive Art: An Argument for Immoralism in the Philosophy of Art.” *Philosophical Quarterly* 71 (3): 481–501. https://doi.org/10.1093/pq/pqaa069.

Liberto, Hallie. 2014. “Exploitation and the Vulnerability Clause.” *Ethical Theory and Moral Practice* 17: 619–29.

Light, Andrew. 2000. "Ecological Restoration and the Culture of Nature." In *Restoring Nature*, edited by Paul H. Gobster and Bruce R. Hull, 49–70. Washington, DC: Island Press.

Lim, Chong-Ming. 2020. "Vandalizing Tainted Commemorations." *Philosophy and Public Affairs* 48 (2): 185–216. https://doi.org/10.1111/papa.12162.

Lindholm, Charles. 2008. *Culture and Authenticity*. Hoboken, NJ: Wiley.

Lindsay, Peter. 2012. "Can We Own the Past? Cultural Artifacts as Public Goods." *Critical Review of International Social and Political Philosophy* 15 (1): 1–17.

Loo, Clement. 2014. "Towards a More Participative Definition of Food Justice." *Journal of Agricultural and Environmental Ethics* 27 (5): 787–809. https://doi.org/10.1007/s10806-014-9490-2.

Lowenthal, David. 1985. *The Past Is a Foreign Country*. Cambridge: Cambridge University Press.

Mantel, Hilary. 2020. *The Mirror and the Light*. New York: Henry Holt.

Manzali, Maira al. 2016. "Palmyra and the Political History of Archaeology in Syria: From Colonialists to Nationalists." *Mangal Media*, October 2, 2016.

Mapes, Lynda V. 2017. *Witness Tree: Seasons of Change with a Century-Old Oak*. New York: Bloomsbury.

Martin, John N. 1979. "The Concept of the Irreplaceable." *Environmental Ethics* 1 (1): 31–48.

Matarrita-Cascante, D., A. Sene-Harper, and L. Ruyle. 2019. "A Holistic Framework for Participatory Conservation Approaches." *International Journal of Sustainable Development and World Ecology* 26 (6): 484–94. https://doi.org/10.1080/13504509.2019.1619105.

Matarrita-Cascante, D., A. Sene-Harper, and L. Ruyle. 2019. "A Holistic Framework for Participatory Conservation Approaches." *International Journal of Sustainable Development and World Ecology* 26 (6): 484–94. https://doi.org/10.1080/13504509.2019.1619105.

Matherne, Samantha. 2022. "Aesthetic Humility: A Kantian Model." *Mind*, 132 (26): 452–78. https://doi.org/10.1093/mind/fzac010.

Matthes, Erich Hatala. 2013. "History, Value, and Irreplaceability." *Ethics* 124 (1): 35–64.

Matthes, Erich Hatala. 2015. "Impersonal Value, Universal Value, and the Scope of Cultural Heritage." *Ethics* 125 (4): 999–1027.

Matthes, Erich Hatala. 2016. "Cultural Appropriation without Cultural Essentialism?" *Social Theory and Practice* 42 (2): 343–66.

Matthes, Erich Hatala. 2017a. "Digital Replicas Are Not Soulless—They Help Us Engage with Art." *Apollo Magazine*. March 23, 2017. //www.apollo-magazine.com/digital-replicas-3d-printing-original-artworks/.

Matthes, Erich Hatala. 2017b. "Repatriation and the Radical Redistribution of Art." *Ergo* 4 (32): 931–53.

Matthes, Erich Hatala. 2017c. "Palmyra's Ruins Can Rebuild Our Relationship with History." *Aeon Magazine*.

Matthes, Erich Hatala. 2018a. "Authenticity and the Aesthetic Experience of History." *Analysis* 78 (4): 649–57.

Matthes, Erich Hatala. 2018b. "Who Owns Up to the Past? Heritage and Historical Injustice." *Journal of the American Philosophical Association* 4 (1): 87–104.

Matthes, Erich Hatala. 2019a. "Cultural Appropriation and Oppression." *Philosophical Studies* 176 (4): 1003–13.

Matthes, Erich Hatala. 2019b. "Environmental Heritage and the Ruins of the Future." In *Philosophical Perspectives on Ruins, Monuments, and Memorials*, edited by Jeanette Bicknell, Jennifer Judkins, and Carolyn Korsmeyer, 175–86. New York: Routledge.

Matthes, Erich Hatala. 2020. "Portraits of the Landscape." *In Portraits and Philosophy*, edited by Hans Maes. New York: Routledge.

Matthes, Erich Hatala. Forthcoming. "Intrinsic and Universal Value in Heritage Ethics." In *The Routledge Handbook of Heritage Ethics*, edited by Andreas Pantazatos, Tracy Ireland, John Schofield, and Rouran Zhang. New York: Routledge.

McGowan, Mary Kate. 2009. "Oppressive Speech." *Australasian Journal of Philosophy* 87 (3): 389–407.

McKenzie, A. D. 2017. "G7 Takes Unprecedented Move to Protect Cultural Heritage." *IDN*, April 24, 2017.

McShane, Katie. 2007. "Why Environmental Ethicists Shouldn't Give Up on Intrinsic Value." *Environmental Ethics* 29 (43): 43–61.

McShane, Katie. 2012. "Some Challenges for Narrative Accounts of Value." *Ethics and the Environment* 17 (1): 45–69.

Meissner, Shelbi Nahwilet. 2018. "The Moral Fabric of Linguicide: Un-Weaving Trauma Narratives and Dependency Relationships in Indigenous Language Reclamation." *Journal of Global Ethics* 14 (2): 266–76. https://doi.org/10.1080/17449626.2018.1516691.

Meissner, Shelbi Nahwilet. 2020. "Reclaiming Rainmaking from Damming Epistemologies: Indigenous Resistance to Settler Colonial Contributory Injustice." *Environmental Ethics* 42 (4): 353–72. https://doi.org/10.5840/enviroethics202042433.

Merryman, John Henry. 1986. "Two Ways of Thinking about Cultural Property." *American Journal of International Law* 80 (4): 831–53.

Merryman, John Henry. 1994. "The Nation and the Object." *International Journal of Cultural Property* 3: 61–76.

Meskell, Lynn. 2011. *The Nature of Heritage: The New South Africa*. Hoboken, NJ: Wiley-Blackwell.

Meskell, Lynn. 2018. *A Future in Ruins: UNESCO, World Heritage, and the Dream of Peace*. Oxford: Oxford University Press.

Miller, David. 2004. "Justice, Democracy and Public Goods." In *Justice and Democracy: Essays for Brian Barry*, edited by Keith Dowding, Robert E. Goodin, and Carole Pateman, 127–49. Cambridge: Cambridge University Press.

Mingon, McArthur, and John Sutton. 2021. "Why Robots Can't Haka: Skilled Performance and Embodied Knowledge in the Māori Haka." *Synthese* 199 (1–2): 4337–65. https://doi.org/10.1007/s11229-020-02981-w.

Molesworth, William, ed. 1994. *The Collected Works of Thomas Hobbes*. Vol. 1. London: Routledge.

Moore, G. E. 1903. *Principia Ethica*. Cambridge: Cambridge University Press.

Paola, Marcello di, and Serena Ciccarelli. 2022. "The Disorienting Aesthetics of Mashed-Up Anthropocene Environments." *Environmental Values* 31 (1): 85–106. https://doi.org/10.3197/096327121x16141642287791.

Peters, Renata F., Iris L. F. den Boer, Jessica S. Johnson, and Susanna Pancaldo, eds. 2020. *Heritage Conservation and Social Engagement*. London: UCL Press. https://www.jstor.org/stable/j.ctv13xps1g.

Polite, Brandon. 2019. "Shared Musical Experiences." *British Journal of Aesthetics* 59 (4): 429–47. https://doi.org/10.1093/aesthj/ayz024.

Quiroz, Lilly. 2022. "The Activist Who Threw Soup on a van Gogh Says It's the Planet That's Being Destroyed." *NPR*, November 1, 2022, sec. Climate. https://www.npr.org/2022/11/01/1133041550/the-activist-who-threw-soup-on-a-van-gogh-explains-why-they-did-it.

Rafii, Raha. 2023. "Digitizing Manuscripts and the Politics of Extraction." In *Variant Scholarship: Ancient Texts in Modern Contexts*, edited by Neal Brodie, Morag M. Kersel, and Josephine Munch Rasmussen, 235–48. Leiden: Sidestone Press.

Raz, Joseph. 2001. *Value, Respect, and Attachment*. Cambridge: Cambridge University Press.

Raz, Joseph, Christine M. Korsgaard, Robert B. Pippin, Bernard Arthur Owen Williams, and R. Jay Wallace. 2005. *The Practice of Value. The Berkeley Tanner Lectures*. Oxford: Clarendon Press.

Redman, Samuel J. 2016. *Bone Rooms: From Scientific Racism to Human Prehistory in Museums*. Cambridge, MA: Harvard University Press.

Riegl, Aloïs. 1982. "The Modern Cult of Monuments: Its Character and Its Origin." *Oppositions* 25: 20–51.

Riggle, Nick. 2010. "Street Art: The Transfiguration of the Commonplaces." *Journal of Aesthetics and Art Criticism* 68 (3): 243–57.

Riggle, Nick. 2021. "Convergence, Community, and Force in Aesthetic Discourse." *Ergo: An Open Access Journal of Philosophy* 8 (47): 615–57. https://doi.org/10.3998/ergo.2248.

Riggs, Christina. 2021. *Treasured: How Tutankhamun Shaped a Century*. London: Atlantic Books.

Ríos-Hernández, Marlén. 2022. "Don't Call Us Posers." *Society for US Intellectual History Blog* (blog). September 1, 2022. https://s-usih.org/2022/08/dont-call-us-posers/.

Rogers, Richard A. 2006. "From Cultural Exchange to Transculturation: A Review and Reconceptualization of Cultural Appropriation." *Communication Theory* 16: 474–503.

Saito, Yuriko. 2007. *Everyday Aesthetics*. New York: Oxford University Press.

Saito, Yuriko. 2017. *Aesthetics of the Familiar: Everyday Life and World-Making*. Oxford: Oxford University Press.

Saito, Yuriko. 2022. "The Aesthetics of Repair." In *Conserving Active Matter*, edited by Peter N. Miller and Soon Kai Poh: 100–119. Bard Graduate Center.

Sandoval, Eduardo Benítez, Omprakash Rudhru, and Qi Min Ser. 2016. "The Birth of a New Discipline: Robotology: A First Robotologist Study over a Robot Maori Haka." In *2016 11th ACM/IEEE International Conference on Human-Robot Interaction (HRI)*, 511–12. https://doi.org/10.1109/HRI.2016.7451831.

Sax, Joseph L. 2001. *Playing Darts with a Rembrandt: Public and Private Rights in Cultural Treasures*. Ann Arbor: University of Michigan Press.

Scarbrough, Elizabeth. 2020. "The Ruins of War." In *Philosophical Perspectives on Ruins, Monuments, and Memorials*, edited by Jeanette Bicknell, Jennifer Judkins, and Carolyn Korsmeyer, 228–40. New York: Routledge.

Scarbrough, Elizabeth. 2022. "Lizzo Playing Madison's Flute and Kim Wearing Monroe's Dress Are Not the Same." *Aesthetics for Birds* (blog). December 2, 2022. https://aestheticsforbirds.com/2022/12/02/lizzo-playing-madisons-flute-and-kim-wearing-monroes-dress-are-not-the-same/.

Scheffler, Samuel. 2007. "Immigration and the Significance of Culture." *Philosophy and Public Affairs* 35 (2): 93–125.

Scheffler, Samuel. 2010a. "The Normativity of Tradition." In *Equality and Tradition*, 287–311. New York: Oxford University Press.

Scheffler, Samuel. 2010b. "Valuing." In *Equality and Tradition*, 15–40. New York: Oxford University Press.

Scheffler, Samuel. 2013. *Death and the Afterlife*. Edited by Niko Kolodny. New York: Oxford University Press.

Schulz, Kathryn. 2022. *Lost and Found*. New York: Penguin Random House. https://www.penguinrandomhouse.com/books/589143/lost-and-found-by-kathryn-schulz/.

Schumacher, Thomas L. 2010. "'Façadism' Returns, or the Advent of the 'Duck-Orated Shed.'" *Journal of Architectural Education* 63 (2): 128–37. https://doi.org/10.1111/j.1531-314X.2010.01073.x.

Scott, David A. 2016. *Art: Authenticity, Restoration, Forgery*. Los Angeles: Cotsen Institute of Archaeology Press at UCLA. https://doi.org/10.2307/j.ctvdmwx02.

Scoville, J. Michael. 2013. "Historical Environmental Values." *Environmental Ethics* 35 (1): 7–25.

Sellars, Wilfrid. 1962. "Philosophy and the Scientific Image of Man." In *Frontiers of Science and Philosophy*, edited by Robert Colodny, 35–78. Pittsburgh: University of Pittsburgh Press. http://www.ditext.com/sellars/psim.html.

Sène, Aby L. 2022. "Western Nonprofits Are Trampling over Africans' Rights and Land." *Foreign Policy* (blog). July 1, 2022. https://foreignpolicy.com/2022/07/01/western-nonprofits-african-rights-land/.

Shelby, Tommie. 2002. "Foundations of Black Solidarity: Collective Identity or Common Oppression?" *Ethics* 112 (2): 231–66.

Sher, George. 2023. "The Weight of the Past." *Australasian Journal of Philosophy* 101 (1): 152–64. https://doi.org/10.1080/00048402.2021.1955288.

Shils, Edward. 2006. *Tradition*. Chicago: University of Chicago Press.

Shim, Joy. 2021. "Literary Racial Impersonation." *Ergo: An Open Access Journal of Philosophy* 8 (31): 219–45. https://doi.org/10.3998/ergo.2232.

Shoemaker, David W. 2003. "Caring, Identification, and Agency." *Ethics* 114 (1): 88–118. https://doi.org/10.1086/376718.

Sibley, Frank. 1959. "Aesthetic Concepts." *Philosophical Review* 68 (4): 421–50.

Sibley, Frank. 2001. "Originality and Value." In *Approach to Aesthetics*, edited by John Benson, Berry Redfern, and Jeremy Roxbee Cox, 119–34. Oxford: Oxford University Press.

Siegel, Jonah. 2022. *Overlooking Damage: Art, Display, and Loss in Times of Crisis*. Stanford, CA: Stanford University Press.

Simons, Marlise. 2016. "Damaged by War, Syria's Cultural Sites Rise Anew in France." *New York Times*, December 31, 2016.

Singer, Peter. 2023. "In Defense of the Art-Targeting Climate Activists." *Project Syndicate*. January 3, 2023. https://www.project-syndicate.org/commentary/eco-activist-non-violent-civil-disobedience-justified-by-peter-singer-2023-01.

Sluss, David M., and Blake E. Ashforth. 2007. "Relational Identity and Identification: Defining Ourselves through Work Relationships." *Academy of Management Review* 32 (1): 9–32. https://doi.org/10.5465/amr.2007.23463672.

Smith, Laurajane. 2006. *The Uses of Heritage*. New York: Routledge.

Spellman, Elizabeth V. 2002. *Repair: The Impulse to Restore in a Fragile World*. Boston: Beacon Press.

Stegner, Wallace. 1969. "Wilderness Letter." In *The Sound of Mountain Water*. New York: Doubleday & Company.

Steinbeck, John. 1939. *The Grapes of Wrath*. New York: Penguin Books.

Stewart, Dodai. 2022. "A Secret Society Tied to the Underground Railroad Fights to Save Its Home." *New York Times*, December 20, 2022, sec. New York. https://www.nytimes.com/2022/12/20/nyregion/united-order-of-tents-brooklyn.html.

Stilz, Anna. 2015. "Decolonization and Self-Determination." *Social Philosophy and Policy* 32 (1): 1–24. https://doi.org/10.1017/S0265052515000059.

Strohl, Matthew. 2019. "On Culinary Authenticity." *Journal of Aesthetics and Art Criticism* 77 (2): 157–67. https://doi.org/10.1111/jaac.12631.

Taylor, Joel, and Laura Kate Gibson. 2017. "Digitisation, Digital Interaction and Social Media: Embedded Barriers to Democratic Heritage." *International Journal of Heritage Studies* 23 (5): 408–20. https://doi.org/10.1080/13527258.2016.1171245.

Taylor, Paul C. 2016. *Black Is Beautiful: A Philosophy of Black Aesthetics*. Hoboken, NJ: Wiley-Blackwell.

"Text of the Convention for the Safeguarding of the Intangible Cultural Heritage." 2003. Paris: *UNESCO*. https://ich.unesco.org/en/convention.

"This Pig Wants to Party: Maurice Sendak's Latest." 2011. *Fresh Air*. NPR. https://www.npr.org/2011/09/20/140435330/this-pig-wants-to-party-maurice-sendaks-latest.

Thompson, Ahmir Questlove. 2022. "Collecting Is an Act of Devotion, and Creation." *New York Times*, March 25, 2022, sec. Opinion. https://www.nytimes.com/2022/03/25/opinion/questlove-inspiration-collection.html.

Thompson, Erin L. 2022. *Smashing Statues: The Rise and Fall of America's Public Monuments*. New York: W. W. Norton.

Thompson, Janna. 2000. "Environment as Cultural Heritage." *Environmental Ethics* 22: 241–58.

Toolis, Erin E. 2017. "Theorizing Critical Placemaking as a Tool for Reclaiming Public Space." *American Journal of Community Psychology* 59: 184–99.

Tuan, Yi-Fu. 1977. *Space and Place: The Perspective of Experience*. Minneapolis: University of Minnesota Press.

Turner, James Morton. 2013. *The Promise of Wilderness*. Seattle: University of Washington Press. https://uwapress.uw.edu/book/9780295993300/the-promise-of-wilderness.

Turner, Lauren. 2016. "Palmyra's Arch of Triumph Recreated in London." *BBC News*, April 19, 2016.

University College London. 2023. "Preserving Endangered Languages as 3D Shapes." January 26, 2023. https://phys.org/news/2023-01-endangered-languages-3d.html.

Valdman, Mikhail. 2009. "A Theory of Wrongful Exploitation." *Philosophers' Imprint* 9 (6): 1–14.

Velleman, J. David. 2002. "Identification and Identity." In *Contours of Agency: Essays on Themes from Harry Frankfurt*, edited by Sarah Buss and Lee Overton, 91–123. Cambridge, MA: MIT Press.

Venture, Tanya, Caitlin DeSilvey, Bryony Onciul, and Hannah Fluck. 2021. "Articulating Loss: A Thematic Framework for Understanding Coastal Heritage Transformations." *The Historic Environment: Policy & Practice* 12 (3–4): 395–417. https://doi.org/10.1080/17567505.2021.1944567

Vogel, Steven. 2015. *Thinking like a Mall: Environmental Philosophy after the End of Nature*. Cambridge, MA: MIT Press. https://doi.org/10.7551/mitpress/9780262029100.001.0001.

Walter, Nigel. 2021. *Narrative Theory in Conservation: Change and Living Buildings*. New York: Routledge.

Warren, Karen J. 1989. "A Philosophical Perspective on the Ethics and Resolution of Cultural Property Issues." In *The Ethics of Collecting Cultural Property*, edited by Phyllis Mauch Messenger, 1–25. Albuquerque: University of New Mexico Press.

Watene, Krushil. 2016. "Valuing Nature: Māori Philosophy and the Capability Approach." *Oxford Development Studies* 44 (3): 287–96.

Whyte, Kyle Powys. 2016. "Food Justice and Collective Food Relations." In *The Ethics of Food: An Introductory Text with Readings*, edited by Anne Barnhill, Mark Budolfson, and Tyler Doggett, 122–35. New York: Oxford University Press.

Whyte, Kyle Powys. 2017. "Our Ancestors' Dystopia Now: Indigenous Conservation and the Anthropocene." In *The Routledge Companion to the Environmental Humanities*, edited by Ursula Heise, Jon Christensen, and Michelle Niemann, 222–31. New York: Routledge.

Whyte, Kyle Powys. 2018. "Settler Colonialism, Ecology, and Environmental Injustice." *Environment and Society: Advances in Research* 9: 125–44.

Williams, Bernard. 1981. *Moral Luck*. Cambridge: Cambridge University Press.

Wonderly, Monique Lisa. 2016. "On Being Attached." *Philosophical Studies* 173 (1): 223–42. https://doi.org/10.1007/s11098-015-0487-0.

Wonderly, Monique Lisa. 2021. "Agency and Varieties of Felt Necessity." *Ethics* 132 (1): 155–79. https://doi.org/10.1086/715290.

Wylie, Alison. 2005. "The Promise and Perils of an Ethic of Stewardship." In *Embedding Ethics*, edited by Lynn Meskell and Peter Pels, 47–68. Oxford: Berg.

Yarrow, Thomas. 2018. "Retaining Character: Heritage Conservation and the Logic of Continuity." *Social Anthropology* 26 (3): 330–44. https://doi.org/10.1111/1469-8676.12532.

Yarrow, Thomas. 2019. "How Conservation Matters: Ethnographic Explorations of Historic Building Renovation." *Journal of Material Culture* 24 (1): 3–21. https://doi.org/10.1177/1359183518769111.

Young, Iris Marion. 2011. *Justice and the Politics of Difference*. Princeton, NJ: Princeton University Press.

Young, James O. 2005. "Profound Offense and Cultural Appropriation." *Journal of Aesthetics and Art Criticism* 63 (2): 135–46.

Young, James O. 2008. *Cultural Appropriation in the Arts*. Oxford: Blackwell.

Young, James O., and Conrad G. Brunk. 2012. *The Ethics of Cultural Appropriation*. Oxford: Blackwell.

Ypi, Lea. 2017. "Structural Injustice and the Place of Attachment." *Journal of Practical Ethics* 5 (1): 1–21.

Yuko, Elizabeth. 2021. "The Terrifying Rise of Haunted Tourism." *Bloomberg.com*, October 28, 2021. https://www.bloomberg.com/news/features/2021-10-28/when-ghost-hunters-become-historic-preservationists.

Ziff, Bruce, and Pratima V. Rao. 1997. *Borrowed Power*. New Brunswick, NJ: Rutgers University Press.

INDEX

For the benefit of digital users, indexed terms that span two pages (e.g., 52–53) may, on occasion, appear on only one of those pages.

activism, 90–92, 148
aesthetics, 4–5, 8–9, 36–38, 48–49, 56, 67–68, 77–78, 160–62
agency, 71–73, 84–87, 93–95, 127–28, 152–54, 161–62
Anishinaabe, 41–42, 117–19
Appiah, K. Anthony, 123–24, 159, 161–62
appropriation, 127–54
architecture, 35–36, 39–40, 78–79, 105, 107–8, 110–11, 113–15, 157–58, 171–72
attachments, 36–37, 52, 66–67, 85–90, 120, 124–25, 151–52
attitudes, 10–12, 34–35, 39–41, 46–48, 53–59, 72–73, 88–91, 112–13, 155–56, 166
aura, 24–26, 35–38, 71–74, 116
authenticity, 17–18, 26–35, 37–38, 42, 55–56, 75, 83–87, 95–96, 99, 105–6, 111–16, 155–56

Brady, Emily, 36–37, 77–78, 79–80

character, 8–9, 35–41, 67, 71–74, 77–84, 134–35, 166, 168–71, 175
climate change, 1, 91–92, 94–95, 111–12
Cohen, G. A., 22, 162–63
colonialism, 2–3, 55–56, 81, 83–84, 88–89, 90–91, 95–96, 131, 135–36, 140, 152–53, 156–57
community, 1–3, 29–31, 38–39, 40–41, 55–56, 58–59, 66–67, 76, 82–83, 89–90, 94–96, 106–7, 110–11, 115–23, 127–29, 132–37, 139–49, 160–62, 169, 170–71
Confederate monuments, 38–39, 88, 90–91, 148
conservation
  of architecture, 35–36, 39–40, 78–79, 105, 107–8, 110–11, 113–15, 157–58, 171–72
  of art and artifacts, 2–3, 5–8, 24, 45–49, 67–68, 103–5
  of cuisine, 30–33, 99–100, 101–2, 111, 117–18, 128–29, 170–71, 173–74, 175
  of digital media, 111–18, 150–51

conservation (*cont.*)
of environments, 23–24, 29–30, 34–35, 39–40, 68–69, 74, 77–81, 91–92, 99–100, 105–6, 107–9, 133–34, 135–36, 140, 156, 163–64
of heirlooms, 16, 20–22, 34–35, 48–49, 51–52, 106, 107–8, 123–24, 144–45, 152–53, 171–72
of identity, 17–18, 19–21, 33–38, 44–45, 49–50, 53, 59–69, 74–75, 85–88, 96–98, 100–3, 107–8, 109–10, 114–15, 118–25, 127–29, 147, 148–49, 152–53, 156, 170–71, 173–77
of languages, 1–2, 59–60, 65–67, 95–97, 110–12, 116–17, 118–19, 135–37
of meaning, 18, 21–22, 34, 43–44, 53, 56, 59–60, 62–63, 71–72, 74–76, 90–91, 93–98, 100–1, 103–26, 138–39, 141–42, 159, 164–65, 167, 171, 176–77
of memories, 17–18, 20–22, 70
of nature, 9–10, 29–30, 34–35, 58–59, 94–95, 163–64
of species, 4–5, 50–51, 65, 68–69, 99, 110–11, 117–18, 124, 161–62
of traditions, 1–2, 18, 31, 34–35, 40–42, 43–44, 59–60, 75, 87–88, 95–96, 97, 107–8, 121–22, 125, 140–41, 159, 161–62, 171–75
conservation vs. preservation, 9–10
continuity, 41–42, 58–59, 76–84, 121, 175
control, 44, 55–56, 72–73, 87, 93–96, 117, 125, 127–30, 132–33, 135–36, 144, 146, 149–54, 156–57, 165–37
cuisine, 30–33, 99–100, 101–2, 111, 117–18, 128–29, 170–71, 173–74, 175
cultural appropriation. *See* appropriation
culture, 33, 48–49, 67–68, 84, 92–97, 103, 111–12, 128–29, 132–37, 148–49, 164–65, 170–72

damage, 71, 79–80, 86–87, 91–92, 93–94, 170
destruction, 4, 24–25, 29, 53–57, 70, 75, 79, 90–95, 108–9, 111–12, 125, 128–29, 138, 152–53
digital media, 111–12, 117, 150–51
digital methods of conservation, 24–25, 111–18, 121
discontinuity. *See* continuity
disrespect. *See* respect
domination, 14–15, 34–35, 81, 95–96, 129, 152–53, 164–65
Dominguez Rubio, Fernando, 103–4, 105–6, 138

environment, 23–24, 29–30, 34–35, 39–40, 68–69, 74, 77–81, 91–92, 99–100, 105–6, 107–9, 133–34, 135–36, 140, 156, 163–64
Eurocentrism, 92–93, 156, 164–66
exemplification, 137–42
expectations, 26–28, 31–33, 37–41, 104–5

fakes, 34–35, 55–56, 112–13, 155–56
familiarity. *See* strangeness
felt needs, 86–88, 120
forgeries, 26–28

heirlooms, 16, 20–22, 34–35, 48–49, 51–52, 106, 107–8, 123–24, 144–45, 152–53, 171–72
heritage, 12, 20–21, 34, 38–39, 55–56, 58–60, 75, 78–79, 92–93, 103, 111–13, 115, 155–65, 169–89

ice cream, 8–9, 12–13
identity
  collective, 33–34, 40–42, 44–45, 58–59, 65, 76–77, 81–84, 87–94, 100–2, 118–25, 127–29, 132, 147, 148–49, 152–53, 156, 170–71
  individual, 17–18, 19–21, 33–38, 44–45, 49–50, 53, 59–69, 74–75, 85–88, 96–98, 100–3, 107–8, 109–10, 114–15, 118–25, 127–29, 156, 170–71, 173–77
Indigenous communities, impacts on, 2–3, 29–30, 41–42, 55–56, 76–77, 80, 133–34, 135–37, 163–64
instrumental value. *See* value
integrity, 76–78, 83–84, 106, 148–49
intrinsic value. *See* value
Irvin, Sherri, 137–39

Jeffers, Chike, 95–96, 140–41

Korsmeyer, Carolyn, 23–24, 25–28, 71

languages, 1–2, 59–60, 65–67, 95–97, 110–12, 116–17, 118–19, 135–37
loss, 1, 4–5, 11–12, 14–15, 20–22, 29, 41–42, 44–45, 50, 51, 60–62, 65–68, 70–73, 75, 79–80, 87–88, 95–98, 171, 174

magic, 23, 26–28, 67
Māori culture, 87–88, 121–22
meaning, 18, 21–22, 34, 43–44, 53, 56, 59–60, 62–63, 71–72, 74–76, 90–91, 93–98, 100–1, 103–26, 138–39, 141–42, 159, 164–65, 167, 171, 176–77
Meissner, Shelbi Nahwilet, 135–36, 140–41
memory, 17–18, 20–22, 70
Monet, Claude, 5–6, 45–47, 51
Muñoz Viñas, Salvador, 12–13, 29–30, 32–33
museums, 1–2, 5–7, 17–18, 26–28, 52, 55–56, 75, 91–92, 99–100, 106–10, 129, 132–34, 140, 148–50, 155–57, 160, 162–63
Mustangs, 85–87, 120

narrative, 71–83, 88–89, 112–17, 135–36
national parks, 1–2, 29–30, 88–90, 133–34, 157–58, 163–64, 166–67, 169–71
nature, 9–10, 29–30, 34–35, 58–59, 94–95, 163–64

Palmyra, Ancient, 24–25, 92–93, 112–14

parenting, 60–62, 68–69, 72–73
participation, 97–98, 116–26, 127–28, 131–32, 134–37, 139–42, 152–54, 156–57
Potawatomi, 65–67
pottery, 18, 52, 55–59, 155–56
power, 14–15, 33–34, 38–39, 55–56, 81, 93–94, 125, 134–36, 137, 139–42, 151, 153–54, 156–57, 164–65, 176–77
preservation. *See* conservation
proliferation problem, the, 50–51, 162–63

Raz, Joseph, 57, 125–26
relationships, 11–12, 34–37, 58–60, 74–75, 85–87, 102–3, 105–6, 107–10, 117–20, 123–24, 135–36, 148–49
repatriation, 134, 148–49, 156–57
replicas, 24–28, 32–33, 50–51, 114–87
respect, 44–45, 46–48, 53–59, 155–56, 167
ruins, 23–24, 90–91, 92–93, 105, 113–14

Ship of Theseus, the, 19–20, 36
social media, 1–2, 55–56, 155–56
species, 4–5, 50–51, 65, 68–69, 99, 110–11, 117–18, 124, 161–62
stewardship, 68–69, 134, 151–54
strangeness, 77–78, 79–80, 82, 175

tacos, 31–33, 99–100, 173–74
tradition, 1–2, 18, 31, 34–35, 40–42, 43–44, 59–60, 75, 87–88, 95–96, 97, 107–8, 121–22, 125, 140–41, 159, 161–62, 171–75
transgression, 33, 56–57, 90–91
trees, 1–2, 23–24, 78, 79–80

UNESCO, 58–59, 75, 92–93, 111–12, 135–36, 157–63
universal value. *See* value

value
  age, 24–25, 113–14
  historical, 22–28, 50–52, 58–59, 73–74, 112–15
  instrumental, 45–49, 65–66
  intrinsic, 44–53, 65–66, 87
  universal, 123–24, 155–72

Whyte, Kyle Powys, 41–42, 76, 83–84, 117–18
World Heritage. *See* UNESCO

Yarrow, Thomas, 78–79, 102–3, 114–15, 122–23

Zuni, 53–55, 75, 132–33